H 5

A

Malcolm McIntyre

HOME EXTENSIONS

A Householder's Guide

Stanley Paul
London

To Avis

Stanley Paul & Co Ltd
3 Fitzroy Square, London W1

An imprint of the Hutchinson Publishing Group

London Melbourne Sydney Auckland
Wellington Johannesburg and agencies
throughout the world

First published 1976
© Malcolm McIntyre 1976

Printed in Great Britain by offset litho by
The Anchor Press Ltd and bound by
Wm Brendon & Son Ltd
both of Tiptree, Essex

SBN 0 09 126880 X (cased)
        0 09 126881 8 (paperback)

Contents

While dimensions in this book are quoted in both
metric and imperial units, the metric has been rounded
off to the nearest maximum or minimum measurement
in line with Building Regulations 1972. It should also
be stressed that measurements in this book are for
guidance and not intended as a basis for final detailed
drawings.

The author and publishers acknowledge the following
for permission to reproduce copyright photographs:
Blacknell, Banbury, Roomaloft, Crescourt.

Acknowledgements

The inspiration of family and friends has been the springboard for this book, but I owe a special debt of gratitude to Winefride Jackson of the *Sunday Telegraph* (where it all began) for her wise counselling over five years and agreement to the reproduction of the case histories I collected in that time, to her staff and colleagues, present and past, particularly Victoria Reilly; to John Martin of Roomaloft, who set me on the path; to Mr and Mrs Tony Longstaffe of Sunnyside Home Extensions; to Bob Tattersall, Editor of *Homemaker* magazine, for his encouragement and invaluable contribution on do-it-yourself; to Clifford Jones and David Papworth, artists of camera and pen; to Roddy Bloomfield of Stanley Paul, for his friendly guidance throughout; and to some very dear people, my wife Avis, who pounded the keys so devotedly in producing the typescript, to my daughter, Claire, who counted the words so painstakingly and to my son, Paul, and daughter, Kate, for their infectious enthusiasm in the entire project.

Introduction

The traditional pattern of home-making in this country, where the Englishman's house is considered his castle, has been for people to uproot and move to larger homes to make room for a growing family and to meet the demands of improved living standards as the prospects and salary of the breadwinner increase over the years. Financial experts continually remind us that owning a home is the best – as well as the biggest – investment that the average family man will make in his lifetime. It is in the family's interest, therefore, to make that investment grow so that a profitable return can be made when it comes to selling the house, say at retirement.

We are living, however, in a time of considerable economic and social pressure, caused on the one hand by the ever-increasing call for home ownership and on the other by the bleak reality that not enough houses are being put up each year to satisfy the demand – and that demand grows more intense as inflation bites more and more deeply into the family's savings and investments. What is likely to help relieve this pressure?

It is worth recalling that Britain is still almost unique in its vast number of freehold properties compared to leasehold, and this clear indication of our owner-occupier heritage suggests that the depressed state of the housing market in this country cannot be allowed to last for very much longer.

A great deal has been talked about the potential value of schemes aimed at helping more people towards home ownership, especially the young. The 'low-start' mortgage scheme is just one. Another idea which has been studied by Housing Ministry officials and property advisers is for mini-homes, built with only one bedroom and a combined living/dining room, plus kitchen and bathroom, but sold complete with plans for extensions.

The advantage is that such houses would grow with the families and allow them to finance the additional rooms when they could afford them. The disadvantage is that rising building costs would add considerably to the eventual total cost of the house. Building societies might consider the idea as a way of spreading a loan and giving 'fair shares for all' by providing, say, £5000 as a first payment, a further £2500 after three to five years and a final £2500 after another three to five years. The idea highlights the ingenuity of planners in finding an answer to the widespread need for a home that grows and grows.

An increasing amount of cash has been flowing back into building societies; however, it is bound to take some time before the market comes anywhere near its previous peak and we can assuredly look forward to a welcome, but still fairly restricted, development of the housing market.

With all these various factors playing their part, how do we ensure that, once we have the bricks and mortar asset, we make it grow? By extending: adding on something which at once provides you with a bigger and better home and, at the same time, increases its value.

Extending a house usually costs less than moving: it obviates the need to uproot from schools and an established social circle (often the worst wrench in moving on), yet it takes advantage of limited money supply and meets a pressing need for more space as the family grows up, and expands, or there is the additional responsibility of a grandparent to care for.

A further strategic point is that, while the price of second-hand houses has dropped or remained fairly static, land prices have continued to rise, so that many people considering moving to a new larger home from one they purchased perhaps three or four years ago, are finding they can consider themselves fortunate to make any real 'profit' on the sale to put towards the new house. Replacement costs – the price it would be to have the same house built on the same piece of land – have risen above the second-hand market value, too. All this has put moving to a bigger home beyond the reach of a great number of people.

Home extensions, again, provide the answer.

The aim of this book is to introduce the newcomer to the ways and means of extending a home. It spares the indigestible details of such technicalities as solid floor construction and soil pipe fitting regulations; instead, it concentrates on lending an informative and guiding hand through an operation that at best can never be a total joy, but at worst can be a veritable nightmare. It is written, not as a consumerist text, rather as the end product of five years of writing about a subject which, from its simple and creative beginnings, has grown to be a thriving industry serving the highly commendable cause of making a home a better place to live in.

Just imagine the immeasurable value of adding on a room for the children, who, instead of encroaching on your daily entitlement to privacy, have their own to enjoy in an extra room in the house. An investment in a home extension is often far more critical than that taken in moving house. The decision to move on may well simply be taken on the desire to find a different style of house with an extra bedroom, while adding on to an existing home often involves delicate questions of how to retain the character of the property and how to maintain privacy where access to a playroom may well be from your living room.

Let's face it, moving house is an upheaval even in the best regulated circles. It has been estimated that the cost alone may be more than 5 per cent of the value of your previous home. That doesn't include the almost incalculable cost of new carpets and curtains, new fittings and all the hundred and one things that you want to convert to your own taste and familiar use shortly after moving in.

The questions that you face in considering whether to move on or stay put and extend are broadly divided into two areas. First, can you afford to? Secondly, is it going totally to disrupt family and social life? Will it mean upsetting the children by moving them from a school they have just got accustomed to and seem to be gaining confidence in? What about all your friends of long standing at the tennis or bowls club or the neighbour who can always be relied on to babysit at a moment's notice or call in with shopping when you are ill? There's a widowed mother who enjoys her independence and lives close by, but would be upset if you went much further afield. Then there's the greengrocer who is able to lay his hands on the best border plants you can buy at the price. Why not extend?

Any form of home improvement should be regarded as an investment as well as something that enhances the style in which you live. To some extent, the economics of a major improvement, like an extension to the house, depend on the area and particular district in which you reside. You will probably have a shrewd idea of what your house might fetch on the open market, especially if you are able to compare it with similar properties in the neighbourhood. The value of adding another room or two should be directly related to whether you can recover the total cost and profit from the enterprise. A sun room at the rear of the house can be a distinct selling advantage when it features on an estate agent's particulars.

Another way to assess the economic merit of adding a room is to relate the number of bedrooms to the bathing facilities, usually all tucked away in one of the smallest rooms in the entire house. More and more people today are seeking two bathrooms, or certainly one major bathroom and a shower room.

It is obviously not advisable to add £10 000 worth of extension to a house which is in a clearly defined £20 000 property district. The buyer who can afford all those extras is more likely to go to an area which already contains houses with that kind of accommodation.

Young couples attempting to buy their first home are discovering how extremely difficult it is, first to scrape together the necessary deposit and then to stretch their income to repay the loan from a building society or local authority. Their best chance is often to purchase a tiny, terraced house with two bedrooms and two rooms downstairs, perhaps with no bathroom. Yet if they can establish that first foothold – the most critical move in an owner-occupier's life – then there is the target of extending the house to provide a bathroom, new kitchen or third bedroom in the attic.

From small beginnings, larger homes grow!

The individually built house which has its own special character offers the greatest potential for extension, provided of course that the additions do not detract too markedly from the garden and retain the appearance and feeling of character that distinguishes the property from other neighbouring houses and makes it a unique selling proposition.

Sometimes the addition of one large room instead of two smaller ones is a better choice because it can double for parties and a games or children's room.

Always plan an extension as an integral part of the house. For instance, the addition of a room leading directly on from an existing dining room may enable you to knock down the dividing wall between that room and the front living room to create one really big living area. A hall could be opened up to give a larger dining area, with a fitted cupboard inside the front door taking coats, umbrellas, golf bags and the like.

If your present house has an integral garage, this might profitably be converted into a third reception room or leisure room for the children, while a new garage, either on its own or with a room above, is built on at the side of the house. The cost of turning a garage into a habitable room is normally comparatively low: it involves the floating of a new floor, plastering of walls, the provision of insulation and the building of a front wall to replace the double doors. With the basic structure already there, it is really only the shell that requires attention to turn it into a natural, ready-made extension to the house.

Do explore all the alternatives, taking into account the main purpose for which the room or rooms are to be used. A room in the loft will cost more than a side extension, but it may be a better investment in the final analysis.

In the current economic climate the caution 'Don't overdo it' might seem irrelevant, but a home extension has a certain time limit on its usefulness to you. The Americans invented planned obsolescence, but you cannot sell off an extension immediately it has served its purpose.

In these initial planning stages, spare a thought for how it can be used when the children have all left home. A room that is ideal this year for a children's bedroom in the loft might serve as a bed-sitter for granny in ten years' time, but would have been far more convenient if it had been a room at the side of the house with its own external door to give her a certain feeling of independence.

Covenants in the title deeds of a house, which are usually kept by your solicitor, sometimes restrict the alteration or extension of a property. They may lay down that permission has to be obtained from the person who first sold the land on which the house was originally built. Most mortgages deeds, too, stipulate that approval of the building society has to be obtained before any modification is made to the house. This is where the searches carried out by a solicitor make their point. They are also indispensable for winkling out details of local planning applications that may, if successful, adversely affect the market value of your newly-enlarged home.

Your original ideas, therefore, could be subject to a few changes before plans for a home extension are finally given the okay.

One thing is sure: a home extension should be rewarding.

1. Different Types

Extensions come in all shapes and sizes. Big ones, small ones, square, rectangular, timber, brick, pre-cast, plain finishes, rendered, flat and pitched roofed. The choice can be bewildering and the decision on exactly how the extension is to be used is the way to narrow down the selection.

Additions

The simplest way is to add on and there are three main types of room: the conservatory/playroom, the living/bedroom and the free-standing rooms sited alongside the house. Building Regulations classify the conservatory room as non-habitable and the living room as habitable.

Every room has to comply with high standards of fire resistance and the habitable room is singled out for special attention. This must have higher standards of insulation and a solid fire-proofed roof at least 2·30m (7ft 6in) high. End panels or walls also have a minimum required thickness. Standard timber extensions, which are enormously variable in overall size and finish, usually conform in one vital respect – safety – and therefore comply with all the current regulations.

Extensions of all shapes and sizes. Here are just a few ideas of how to add on extra living space. The top drawing shows how two ready-made rooms have been put together to make an 'L' shaped extension leaving an open-roofed patio behind the kitchen. The next house has a new room in the loft and again two rooms added on to the side and rear. The little cottage-style bungalow has been rather ingeniously extended at the back to make an extra bedroom and study with its own sun balcony. The large detached house has been extended to create a room over the garage and another in the new loft area.

Within this wide category of standard extensions come ones for the do-it-yourself enthusiast. Sections all ready to assemble are delivered to your home and the main job that has to be done is to prepare a concrete base. The need to get all measurements absolutely spot-on goes almost without saying and you may prefer to leave the task to a local builder or a couple of men who are anxious to earn a little 'on the side' at the weekends. If you feel up to tackling the job then infinite care in preparation before calling in a firm to supply the ready-mixed concrete will pay dividends.

The height of an add-on room varies between 2·15m (7ft) and 2·75m (9ft) and there is a choice of length ranging from around 1·7m (5ft 6in) to well over 6m (20ft). Widths vary from 1·5m (5ft) to about 3m (10ft).

The various nationally known firms that produce prefabricated extensions manufacture a whole range of buildings from garden sheds to games rooms. Some of the leading names offer more than 100 variations to choose from. Their aim is, as far as possible, to provide you with an individually tailored extension to suit your particular requirements.

They promote themselves through the press and a completed coupon will bring you through the post a detailed sales brochure complete with price list and order form. You can also ask for a representative to call. Sending off for a number of brochures for this type of ready-made extension is by far the best way to start. From these you can assess prices and weigh up the pros and cons of just what is included in the 'inclusive price'. These catalogues do vary quite markedly in the details they give you so the watchword is Astuteness. One particular company, for instance, will provide a solid end wall to replace their 'standard' timber side frame, at no extra charge. Others feature guttering as a standard fitting, while several do not, and it is all these necessary extras that are hidden costs until you sit down and work it out in detail.

The biggest advantage of the ready-made unit is its lower cost. If you lay the concrete base and get a neighbour to help you erect it, you can save quite a bit on the total budget. Some firms, not many, will assemble a room for you; most are only interested in selling a

Concrete panels of this extension are in a mock brick finish to blend in with the bricks of the house.

kit of component sections and providing complete instructions with easy-to-follow diagrams on how to erect it.

Delivery is free with some firms, depending in which part of the country you live and how far from their head offices or one of their show sites.

Guarantees need to be looked at with a critical eye. One company, for example, offers a 'world-famous' guarantee for any goods made by them. But their products are sold on an owner erection basis only which begs the question: 'What about damage incurred accidentally while erecting it to the company's own instructions?'

One way over this is to employ a builder to put up the extension and then leave him to negotiate with the manufacturer so that any question of responsibility is accepted by the builder in a written agreement.

In any case, it may be necessary to obtain planning permission before one of these rooms is erected so it is as well to ensure that approval is granted before finally committing yourself to the order. Certain firms will supply drawings of their units for use when applying for planning permission. However, in some cases, they may have to have additional details included. The ideal solution is to get the firm to handle this aspect of the transaction for you.

When working out the total budget, do remember that on top of the price of the room will have to go the cost of any electrical and plumbing work.

If you want to make the extension into a habitable room, under the terms of the Building Regulations, then it must have a ceiling height of 2·3m (7ft 6in) or more and the walls and roofs have to conform to a minimum standard of insulation and fire resistance. The floor, too, has to have a damp-proofing in this case. All these points make a good deal of difference to the final price tag.

Roofs can be in translucent PVC or of more solid construction with insulated chipboard on top of which goes three layers of felt covered with chippings. There's a wide choice, too, in the types of timber doors and windows.

The newest developments are vinyl coated finishes which reduce maintenance to an absolute minimum and a special pre-cast block with high insulative qualities.

Modern garden room that balances well with this older-style house.

Sun room that is equally suitable as a detached room in the garden or adjoining a house.

In the siting of a ground-floor extension the things to avoid, literally, are drains. If a proposed extension covers these, thereby increasing the load on them, they must be reinforced. Although it is possible to leave an existing inspection chamber inside a new building – fitting it with an airtight cover by leaving it so that it can be removed if necessary – the local authority may insist that it is repositioned outside. This could be difficult if you are running close to the boundary and only a professional inspection will provide the answer.

Labels (top drawing): RAFTER, HANGER, PURLIN, CEILING JOIST, BINDER, STRUT

Labels (bottom drawing): NEW CEILING JOISTS, DORMER WINDOW, MUST BE AT LEAST 8ft. 2ins. 2·49m., 7ft 6ins 2·29m, NEW STRUTS FORMING WALLS, NEW FLOOR, NEW FLOOR JOISTS

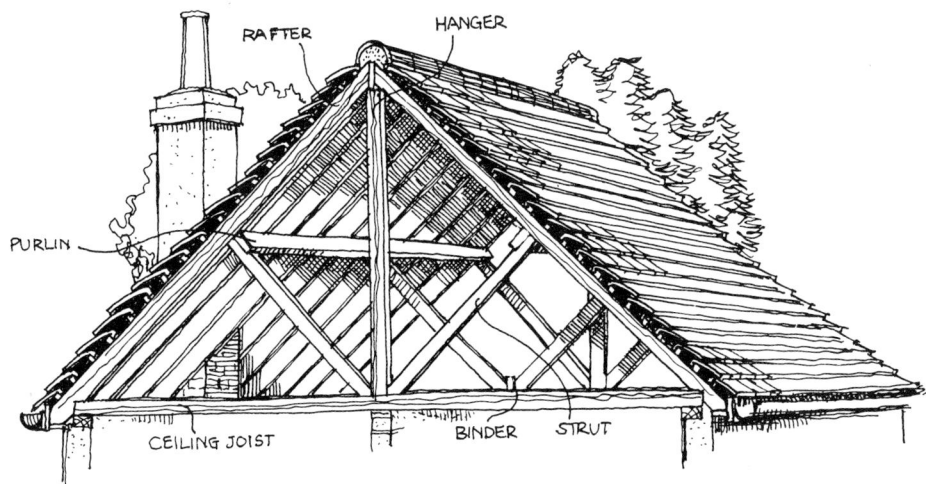

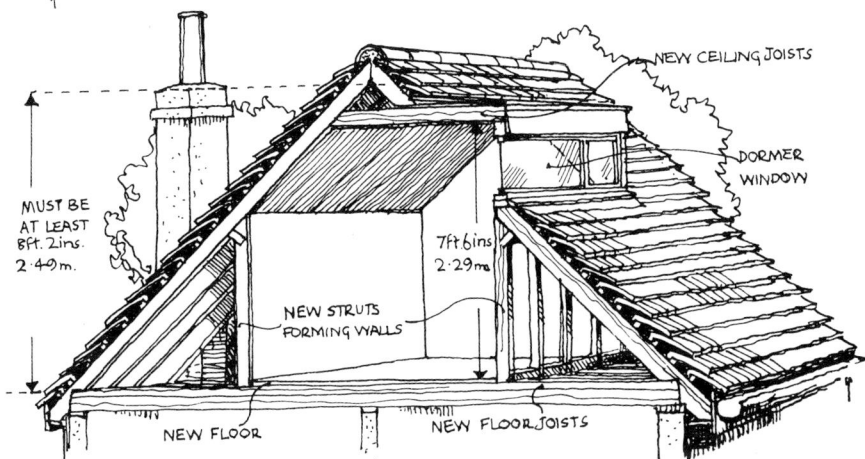

By comparing these two drawings you will see how a typical loft area has been converted to make an extra room. The ceiling joists have been strengthened to take a new floor; a dormer window added to gain extra height and new roof struts put in to carry the weight of the roof.

Lofts

For rooms in the attic it pays to employ a specialist for this form of conversion can be a tricky job, needing very careful planning. This is because the structure of the roof has to be altered, and changing the position of timbers supporting the roof and putting in new supports is a job reserved for the professional.

Indeed, it has been the rapidly growing interest in making use of that largely wasted space in the loft that has led to the increasing number of specialist loft conversion companies now vying with one another for this lucrative part of the homes market.

When you consider that the loft has a potential floor area possibly bigger than any of the rooms in the rest of the house, you can see that it offers perhaps the most interesting and often visually most pleasing way to extend the home.

It has three obvious advantages over the add-on room:

It does not take up garden space.

It makes economic use of wasted space that already exists in the house.

It retains the original character of the property with a dormer window often being the only externally visible clue to the fact that an extension has been carried out.

This last consideration is, to me, an extremely important one. The last thing in the world you want is an addition that looks like an architectural afterthought!

In the planning checklist which follows in Chapter 4, some circumstances which could make a loft conversion an uneconomic proposition, if not an impossibility, are stressed. They are concentrated on the actual shape of the roof. This determines the height of any new room to be created and there is a minimum decreed under Building Regulations: this is not less than 2·3m (7ft 6in) over at least half of the room. Allowing for deeper floor joists, a new floor and the depth of a ceiling, you will need a minimum height of 2·5m (8ft 2in) over at least half the proposed floor area when you come to measure the attic.

If the minimum height is not there, it may be possible to achieve it by installing a dormer window. When the height is there, then a pivot window, which can give more light, is a successful alternative.

As a guide, an attic of some 6m (20ft) long could give you a 3·5m (12ft) square room.

When you take a look in the loft, check on the obstructions there are, like water tanks and the expansion tank for hot water heating systems. These can be moved fairly easily. The ideal place is directly behind one of the new walls with access gained through a small door.

You are certainly better off with an older house when thinking of using the attic. Houses built since about 1965 are likely to have a

Spacious room created by the conversion of an attic.

roof constructed with what are known as trussed rafters. These are made of comparatively thin timber joined with metal plates. The whole thing is built with such inter-related precision that to take one rafter out would mean the possible collapse of the rest of the roof. Thus, you would be faced with a major reconstruction job for a loft conversion.

It is worth seeking the advice of an architect or surveyor, however, if a room in the loft would appear to be the best way of extending your home. With the older-style house, it is far simpler to move supporting timbers without an adverse affect on the stability of the roof and you can gauge this just by looking up into the roof space. Larger and stronger floor joists will have to be put in because the existing joists were only ever intended to carry a ceiling. They will need to be a minimum 18 × 5cm (7 × 2in) replacing the 12·5 × 5cm (5 × 2in) joists already there.

New walls in a loft are often made of timber studding; this is lightweight and easy to fix to the roof timbers. There will be spaces behind these new walls and it is worth considering the possibility of making useful storage cupboards out of them. At least ask for a quote from the builder when costing out the conversion. In a semi-detached or terraced property, the party wall can be used as one end of a loft room. Partition walls at the sides of the room can also partly follow the slope of the eaves and are usually 1·7m (5ft 6in) high.

Every habitable attic room must have adequate ventilation and there must be a certain zone of open space outside the window. If a chimney is less than 2·3m (7ft 6in) away, for example, then its top must be at least 1m (3ft) above the top of the window.

Similar restrictions apply where an existing ventilating pipe from the drains ends outside the window. It either has to be moved or extended above the new window. The normal width of windows is 3m (10ft) to 3·5m (12ft) or 10 per cent of the floor area. Anything bigger would put too much strain on the surrounding roof.

Local authorities often insist on having structural engineers' calculations where major alterations are planned. Where a first-floor ceiling is anything other than lath and plaster, it has to be covered above with high density hardboard. The net effect of this sort of regulation is to improve standards as well as safety.

Access to and from a loft room is a major consideration. If it is to be used as a bedroom, then there has to be a permanent staircase; if the room is simply to be a workroom, study or area where the model railway track can be laid for a budding, or even a life-tempered, enthusiast, then a loft-ladder is acceptable. The various types of staircases and windows are covered more fully later in the book.

The omnipotent Building Regulations, which incidentally have only given serious attention to stairways since 1970, have led to vastly improved standards of construction. The conversion of a loft

A growing family is now accommodated with the addition of this loft room, giving the eldest son his 'independence'.

area of a two-storey house into a room, or rooms, has the effect, of course, of turning the building into a three-storey property. Even stricter rules about fire precaution come into play, therefore. To allow a safe escape route in case of fire, these require the internal stairways and associated hallways and landings in houses of three storeys to be separated from the rest of the house by a structure with a fire resistance of half an hour.

Any sizeable extension will increase the rateable value of your property. The local authority has the responsibility of keeping the valuation officer in touch with improvements to local property which have increased their value. Central heating is seen as an improvement under these terms.

If you receive a form through the post advising that it is proposed to increase your rateable value, you have the right to object if you think you have reasonable grounds. Perhaps a newer house close by, with similar accommodation, is rated lower for some reason. You can use this as an example of why your rate should not be altered, so it is well worth checking on neighbouring values when faced with a threatened increase.

Where you are financing a home extension by adding to an existing mortgage or taking out a second, then the building society will automatically insist on increased insurance cover for the actual structure.

2. The Specialists

With the increasing interest in home extensions has come the growth of firms specializing in this type of work. They are formed by men who, in many cases, have no architectural qualifications, but rely on good management of a team of specialists including surveyors, designers and builders, to get business. By and large, their reputation has been well-earned with a ready reference of completed jobs to prove their success. It's as well to take a check on one or two, at least, of the extensions carried out by a specialist firm before you decide to sign a contract. They should be only too happy to let you talk to a 'satisfied customer'. In fact, a lot of their new business comes from word of mouth recommendation and this is always a sure test of satisfaction. The major extensions to the house illustrated on pages 98–99 are just one project of a number that have been completed by the same firm within one locality.

Some firms specialize in loft conversions. This requires particular skill because it means altering the timber supports for the roof, putting in dormer windows and cutting out a space for a new staircase. As a room or rooms in the attic makes valuable use of wasted space – and doesn't necessitate taking up some of the garden – the specialist loft conversion firm flourishes on this business alone.

Both the all-purpose firm and the loft company will assist in designing the extension so that it meets your visual and functional requirements, advising you at the same time on what is likely or unlikely to be acceptable to the local council.

This is the first stage. The second covers the submission of a contract price for building the extension. At this time it is certainly worth obtaining a competitive quotation. A vital guarantee is that the work will be completed 'at the price quoted'. The continual rise in the cost of building materials makes this a difficult thing to arrive at, but if the firm envisages delays in starting they may either write in a let-out clause, which covers increased prices of materials, or resubmit a total cost nearer the time they are ready to begin work. Here is the time to show your strength in bargaining. Be positive about only committing yourself to a certain sum; let the conversion firm take a normal commercial risk of making provision for rising costs – and even reducing profits for the sake of maintaining a regular flow of cash.

Another important clause is one referring to time. The experienced company with a satisfied workforce and a professional reputation, especially with its own suppliers of materials, should be fairly expert at gauging the amount of time it will take to complete an extension, from the initial planning right through to the final day of handing over. There can be delays in getting planning approval, often because the local planning committee receives so many applications a month, a backlog accumulates. However, a telephone call to the council offices can provide a ready answer to this.

The two-storey extension on pages 77–79 is one of the best (or worst, depending which way you look at it) examples of the need for a commitment on time. The worst kind of delays can be on non-standard equipment, so the full specification provided by the specialist firm or builder needs to be gone through with a fine tooth-comb. All this might sound over-cautious, but experience has taught me that when you are spending that kind of money – anywhere between £3000 and £10 000 upwards – it is unquestionably worth investing some time in ensuring that, as far as possible, costly delays are avoided. In any event, it is worthwhile including a penalty clause in the contract to cover time. This means that over a certain specified period, the builder has to bear the additional labour costs. It is a marvellous incentive and a cheap form of insurance.

When the contract is drawn up, some firms will require a deposit, which is refundable in full if you do not approve their initial drawings or planning permission is refused by the local authority. How strictly this rule is applied varies from one firm to another, but £250 on a £5000 extension when you have not even seen first sketches seems a lot to ask. Others require payments as they go along. These are usually linked to each major stage of the job. I talk about financing an extension later in the book; suffice it at this stage to say that any firm that relies on this method of finance should be able to reflect the advantage it gains with reduced overheads by giving you an extremely competitive price. It is, therefore, a good idea to obtain your second quotation from a well-known local builder who is less likely to ask for stage payments.

The next stage, after design, is the preparation of a full structural survey and the drawing up of plans for approval by the local council. Guarantees on the work carried out vary, but the best is ten years which corresponds with the guarantee on new houses provided by the National House Builders Registration Council.

'While every endeavour will be made at time of construction to achieve the size as shown on the approved plan or order, the company cannot be held responsible for any alterations considered necessary on site due to structural reasons. Should materials supplied or any act or omission in any assembly, erection or repair carried out by us prove, to our satisfaction, to be faulty (provided that such faults be notified in writing within ten years of completion of this order) such materials or defects will be at our option replaced or repaired free of charge: fair wear and tear or damage through misuse excepted. This guarantee becomes effective from the completion date.' Signing this sort of contract does not preclude you from taking action against any offending firm in the courts.

The first sentence of this guarantee immediately provides a 'let-out' clause and allows for additional expense that may become necessary once work has begun. The alteration might be due to miscalculation in the plans or specifications or even to a mistake in the original drawings. If you consider that a contract is too one-sided, then you should insist on changes. In any event, it is well worth while checking a contract with your solicitor.

The specialist firm offers you no more than you can get by going to an architect who in turn will instruct a building firm. The main advantage is that you are only dealing with one person and if there are any problems you've only one person to blame; against this has to be set the fact that you are totally reliant on one person or company and have no professional adviser on your side.

The alternative to the specialist firm is to employ an architect. This is a person who must be registered with the Architects Registration Council of the United Kingdom and has to have academic qualifications. Some firms operate under the guise of professionalism by calling themselves architectural consultants or design consultants. It is advisable to check on their credibility by asking about qualifications or examples of work which they have successfully accomplished. Because a professional qualification is not in itself a guarantee of creative flair and speed of work, the personal recommendation remains one of the best ways of ensuring a job well done.

A local estate agent can often recommend an architect and the local council hold a list of local architects for the public to consult. The Yellow Pages of the telephone directory include a list of registered architects in private practice. An architect will prepare plans, drawn to your brief, and then submit them to the local council for approval. He will instruct a firm of builders and supervise construction as it takes place. This is a role that can also be undertaken by the building surveyor who is equally well qualified to draw up plans and specifications for building work, to advise on building and planning applications and to invite estimates and control building operations. He can also give advice on building legislation, boundary disputes and improvement grants.

Architects and building surveyors charge a fee. In the case of architects this is a percentage of the total cost. The minimum RIBA fee for what is referred to as 'normal service' is:

Job cost £	Rate %
Up to 2499	13
2500–7999	12$\frac{1}{2}$
8000–13999	12
14000–24999	11
25000 and over	10

'Normal service' includes advice on alternative schemes, preparing drawings, submitting these for approval by the local authority, drawing up specifications for obtaining tenders, preparing a contract, giving advice on the employment of a builder, working out a timetable, supervising the work and generally seeing the job through to completion. There is a scale of fees, too, for a partial service, should you not require the 'norm'.

On top of these fees will go out-of-pocket expenses for photocopying of drawings and documents needed for submission to the authorities, local travelling and other office expenses. Chartered surveyors' services are costed out on an *ad valorem* basis – that is to say, they are based on the value of the property concerned – or on the calculation of the value of the time it takes to meet all your instructions. Their fees are based on the RIBA scale.

When you are contemplating a very simple extension – say, enlarging a kitchen or adding on a playroom or study – then it may be easier and less costly to show your sketch to a local builder and ask him to quote for carrying out the work. Get him to call and chat about it on the site and ask to see some previous work of a similar character. There is often a neighbour who has employed a local firm at some earlier time and can vouch for them. The local paper is a good source of names through advertisements and then there are the Yellow Pages, of course.

Approach builders in plenty of time because if they are good they are certainly popular and there will be a waiting list, with anything up to a six-month wait before work starts. Ideally, a major extension should be planned for the spring or summer. Construction is a messy job, however fastidious the builder, and there is nothing more aggravating and depressing than mud or brick-dust-encrusted boots tramping in and out.

The local authority must be advised of certain stages of work in building an extension, for example, when the foundations have been completed, the drains put in or the roof constructed. This is to enable the council's Building Inspector to satisfy himself that all the regulations have been complied with. This responsibility rests with the specialist you've employed to carry out the work. Just ensure that such points as these are all covered. That's why it is just as well to get a solicitor to go over the contract for you.

A builder will estimate the cost of an extension by multiplying either the net floor area or the net cubic content by a rate per square or cubic foot. The rate will depend on his overheads, the sort of work involved and prices for materials.

In dealing with any specialists be firm about what you want; because of their professional standing or reputation they will sometimes be inclined to put forward their own idea on what they think should be done. The ideal will adapt your ideas and put the professional touch to them so that the extension not only looks good but is soundly built to a rigid specification and meets all the regulations governing safety.

3. D-I-Y

No matter what type of home extension you decide on, you will cut the cost of it considerably if you carry out the work yourself. But is it possible for the average person to do this? That depends on three factors: the type of extension, its size, and your skill as a do-it-yourselfer.

The simple sun rooms, which feature so prominently in the catalogues of all the extension manufacturers, have been expressly designed to be put up quickly and speedily – which brings them well into the range of the do-it-yourselfer of only limited abilities. These firms have in recent years also started to offer prefabricated, sectional extensions designed to act as a permanent living room, and erecting these calls for only a little more skill, although perhaps a fair amount more in the way of brawn and muscle.

When we move on to brick-built extensions, then we are getting into the realm of work demanding a lot more knowledge, although small add-on rooms of this kind are within the reach of an experienced do-it-yourselfer. A larger brick building would call for considerable knowledge of the building trade and a lot more skill. And as for the really ambitious extensions – two-storey ones, or rooms above a garage – these are best left to the professional.

Likewise, a room in the loft involves so much interference with the structural timbers of the roof that you could do untold damage to your house if you try to create one without knowing what you are doing. But it is possible to cut costs considerably here by letting a builder, or one of the firms specializing in these conversions, carry out the main body of the work, and yourself apply the finishing touches.

This is not the place to detail exactly what is involved in building an extension yourself. If you buy a sectional building, then along with it will come full instructions for erecting it, and most firms supplying these have technical representatives whose brains you will be able to pick. Should you, as a do-it-yourselfer, be contemplating tackling a brick-built extension, then you would be well advised to consult a building textbook.

However, so that you can have some idea of whether you would feel able to tackle the work, let us take a look at what is involved.

Laying a concrete base

Anyone who has ever put up even a simple sun room will tell you that the most difficult part of the whole operation is laying the concrete base. In fact, a base is necessary no matter what type of extensions you propose to build.

First, you must excavate for the base. If there is just soil to be

A base of hardcore provides the foundation for a ready-made room to be added on to an existing house.

dug up, that will not be too much of a chore with an ordinary garden spade. But the site for your extension may well be occupied at present by some sort of patio. Loose-laid crazy paving or concrete slabs can be dug up with a spade, but if you have concrete there, you would be well advised to use an electric hammer. These can be hired.

You must consult both your supplier and/or your local authority's Building Inspector about how the base should be constructed, but more than likely you will be called on to put a 10cm (4in) layer of concrete on top of 10cm (4in) of well rammed hardcore. A broken-up old concrete terrace will go some way to providing this hardcore, but otherwise consult a builder.

You are going to need an awful lot of concrete for the base of even a tiny extension, and you could easily find yourself having to mix and lay four, or even more, tons of concrete. Don't even think of doing this by hand with a bucket and shovel. Your best plan is to use ready-mixed concrete, for those huge lorries that service the building sites are quite willing to come and deliver down your suburban street. What's more, this saves you not only from the necessity of mixing the concrete; it also means that you do not have to bother working out just how much to order of the various materials (cement, sand, gravel) that go to making up concrete. Just tell the supplier the area and depth of base you require and he will send you the right amount of concrete. The charge for ready-mixed concrete is little more than what it would cost you to buy the materials separately.

If the concrete lorry can get down the side of your house to where you are actually building the base, then the job becomes very easy, for the material can be tipped out just where you want it. By wiggling the chute around, and moving the lorry slightly, the driver can go some way to spreading the stuff out for you. The drivers are, by and large, very obliging, although it probably does no harm to give them a tip.

If the lorry cannot get through to the site of your extension, then the concrete may have to be dumped at the front of your house. Transferring this to where it is wanted is not something you can do by yourself – you are going to need a lot of friends with wheelbarrows. You will have something like two hours to move and lay the concrete before it sets. When you think that a cubic yard of concrete which will cover 15 sq m (sq yd) by 5cm (2in) thick, weighs about 2·03 tonnes (2 tons), you have some idea of the size of the task. However, one man with a barrow should be able to move a cubic yard in an hour, so with friends helping you should be able to cope.

If you feel you cannot, then you can pay extra for the lorry to stand by, gently turning the concrete over in its mixer, and delivering its load to you at a slower pace. However this can add quite a bit to your bill.

An alternative to ordering ready-mixed concrete is to hire a mixer and make the stuff yourself. This will not work out much cheaper – in fact, depending on the quantity you need, it could cost slightly more. But the concrete is prepared in more manageable quantities, and you can work with a mixer on days when the ready-mixed firms might not be prepared to deliver, e.g. Sundays and bank holidays.

No matter how you mix the concrete, you pour it into what is known as a formwork, which is a series of stout boards forming a frame all round the shape of the base, and held in place by stout wooden pegs. Then you tamp the concrete down by means of a stout board, which you hold at one end, while a helper takes the other.

You will have to take some trouble over setting up the formwork, for in most cases it is necessary to build the base to accurate size.

A damp-proof membrane has to be incorporated in the base. The usual method is first of all to lay the main base and tamp the surface down to a rough finish. On top of this surface you put down a membrane, which must link up with the main damp-proof course of the house. This membrane may be either in sheet form (e.g. polythene) or a bituminous liquid that you brush on – consult the supplier and the local authority about exactly what is required. Finally, you lay a thin top surface – a screed as it is known in the trade – and this is trowelled down to a smooth finish suitable to receive the flooring material of your choice – vinyl in tile or sheet form being particularly suitable.

You should leave the base for about a week before you start to do any serious work on it, although it will be possible to walk lightly over it after a day or so. In very hot weather, slow down the drying rate of the concrete by covering it with damp sacking.

Assembling an extension

With the concrete base laid, the assembly of a sectional sun room will be a fairly simple job. It is not too easy to generalize about the exact system of putting these together because the various manufacturers have different methods and processes. However, these rooms consist of a series of panels that, in one way or another, are locked together. Full instructions are supplied. The sections are fairly light, and the whole job is easily managed by two men, or perhaps even a man and wife.

Obviously, there have to be watertight joins where the extension meets the house, and the kit will contain devices for ensuring this. A weather-proof sealing strip is fitted where the extension walls meet those of the house. In the case of the roof, it might be necessary to fit some form of flashing. Are you familiar with this term? It describes a strip of material that used to be lead but in the case of an extension kit is more likely to be some form of plastic, that is fitted into a horizontal course of mortar just above the room (you have to cut into the mortar first) and folded down to seal the joint between roof and wall. Modern flashings are very effective and easy to fit.

The roof of this type of extension is usually in corrugated plastic – a light, easily handled material. You fasten it to a series of cross beams, using the proprietary fixing bolts that the makers of the plastic supply.

Some extensions come ready glazed; with others you have to fit the windows yourself. But in the latter case, the glass should be cut accurately to size, and some manufacturers offer patent plastic devices instead of putty to hold the window in place. So you can see that glazing your extension will not be much of a problem.

Erecting a more permanent sectional room is similar in principle, although the work is a lot more hefty. The sections forming the walls are usually heavier, since a solid roof is called for, and this will take a bit more handling. The manufacturers also claim that assembling these rooms is a two-man job, and they may be right, but most of the do-it-yourself assemblers to whom I have spoken found that a third pair of hands came in very useful.

As you work, you may find that the odd part is missing, or that you come up against snags. In such cases do not hesitate to get in touch with the manufacturer, to get the problem sorted out.

I have spoken to many people – some of them not at all experienced do-it-yourselfers – who have erected sectional buildings, and all of them found that, although they might have come across odd snags on the way, in the end they were able to do a thorough workmanlike job.

A brick-built extension

If you are thinking of constructing a purpose-built extension in brick, then you need a lot of knowledge of building techniques and practices. You will have to dig for foundations and construct them. Lay a base. Walls will have to be built and tied into the existing house wall. Similarly a roof will have to be built, doors and windows fitted. If you do not know how to do these things, then you will have to learn, and this can be done under the supervision of an expert, or you will have to study a manual.

Many do-it-yourselfers have put up this sort of extension, and in some cases it has been their first attempt at any ambitious building work. There is no doubt that this is quite an undertaking for the unskilled man.

A room in the loft

If you poke your head into a loft that has not been converted, you will see a tangle of structural timbers, many of them in the middle of it. These have to be cleared away and compensating supports installed instead, so that there will be enough space for a room of decent size. Furthermore, a dormer window has to be fitted. These are jobs requiring a great deal of building knowledge and a fair amount of skill. Although, undoubtedly, many do-it-yourselfers have tackled them, you really ought not to do so unless you understand fully what you are doing, or can receive expert guidance from someone who does.

With the main structural work out of the way, there are many finishing touches that are well within the scope of even a comparatively inexperienced do-it-yourselfer. Many firms are willing to co-operate with the do-it-yourselfer along these lines. For instance, once the ceiling joists have been suitably strengthened by an expert there is no reason why you should not fix the floor. This will be merely a question of nailing floorboards – or, more likely these days, small sheets of chipboard – in place. Then there is not much difficulty about fitting the walls of your new room. The architect or builder who plans it will, in conjunction with the local authority, have specified how these should be built, and they will probably specify sheets of plaster board fixed to a timber framework, or a system of interlocking wall panels made up basically of plasterboard. You should have no difficulty at all with either of these methods, and the manufacturers of these materials offer instructional leaflets.

The ceiling, too, may also well be of plasterboard, and although the principles of constructing this are similar to those for the walls, the fact that you are working overhead introduces a certain amount of extra strain, and it is up to you whether you feel able to cope with it.

Your room will also need skirting board and architrave round the door opening, if there is one. All that is involved here is cutting the timber to size and nailing it in place.

The electrical work needed will be pretty rudimentary, but the consequences of even a slight error can be very serious here, so do not bother with it unless you know what you are about.

Remember throughout that a do-it-yourselfer's work must comply with the regulations just as surely as that of a professional, and will be subject to just the same scrutiny, so make sure you stick to the plans and consult your local Building Inspector all along the line.

4. Planning Checklist

The simplicity or complexity of work involved in building an extension or adding on a room will determine whether or not you need to call on the services of a professional adviser. As a start, it is worth running through your own checklist of planning points.

What sort of extension?

To what use is the room (or rooms) to be put? Is it simply a question of a little extra space when entertaining friends or holding a party? Or is it to be put to more permanent use, say a granny coming to live in or a fast-growing member of the family who deserves a room of his or her own for the first time? That's the most important consideration.

Having decided on this, the next question is: where? Where is the room to go? Maybe alongside or at the rear of the house, or, perhaps because you've a small garden and the boundary runs too close to allow for any additions, then it's up into the roof to make the best of an inviting area of space.

Where structural alterations are involved the job should obviously go to a professional, but in the case of the add-on room you can sharpen your wits and become a part-time brickie by doing it yourself. Going it alone is described in the previous chapter; suffice it to say at this point that you need a whole lot of courage, determination and time to get the best out of what must be a rewarding – and money-saving – experience.

Incidentally, where an extension, especially the two-storey kind, is going to run within a few inches of the boundary, do remember to have a neighbourly word with the people next door: it can stop a whole lot of trouble and aggravation in piloting your planning application through the sometimes ponderous corridors of the local authorities. Often, because of the large numbers of applications, it can take several months to obtain permission for what might at first seem to be a 'rubber-stamping' job. So it's a positive investment to remove, in advance, any possible stumbling block, because some will certainly be there, even in the best organized extension to the home.

A word with your solicitor at the outset is a worthwhile precaution. He will check on the title deeds of the house to see whether there are any restrictions concerning extensions and will institute a search to see whether any local covenants pose a problem. For example, if you buy additional land on which to build the extension there may be prohibitive clauses on new buildings being erected; this type of hurdle can perhaps be overcome, but it's wise to plan ahead.

The siting of a ground-floor addition needs to be considered carefully. Existing drains will present problems, although these can sometimes be overcome at a price!

The proposed design of an extra room that would bring it up to a boundary wall may have to be modified because of the requirements for fire precautions. For instance, the amount of window space you can have will depend on the distance of the extension from the boundary.

Size

When you start planning an extension in your mind, think ahead to the future when you may need even more space. It could pay today to lay the foundations for a two-storey extension even though at the moment you only need a single-storey one. Costs do not increase *pro rata* with floor area and the cost per unit of usable floor area of a two-storey extension would be much less than for a single addition because the main construction work is already done.

The extra cost of building a larger room increases more slowly than the actual size of the room. For instance, the cost of building a room half as big again as you might originally plan, would only cost some 30 per cent more on the total budget. This is particularly true when you employ a local builder for he has his set overheads for a job, including ordering materials and arranging for his men to be on site over a certain time whatever the size of room. On the other hand, of course, there is simply no point in having a large room for the sake of it for there are heating costs and increased rates to consider apart from the value it is adding to the house.

Building Regulations

The Building Inspector of your relevant local authority takes his lead from the Building Regulations – a set of rules that, although overflowing with Civil Service jargon that will confound the average person, contains all the factors for safety and commonsense necessary to complete an investment that will add value to your home in terms of use as well as money.

Building Regulations, which are being continually updated as new materials come into use and new ideas supersede old, took the place of local authority byelaws in the late 1960s. It is a document 188 pages long with a general section covering the interpretation of the regulations and their application, several more dealing with the fitness of materials to be used, getting the site ready, preventing damp, ensuring adequate fire protection, permitted types of stairways and balustrades, the amount of space necessary outside windows of habitable rooms and, very importantly in today's conservation of heat and energy, a series of regulations on thermal insulation of roofs, walls and floors. In addition there are schedules which are explanatory rules and tables for satisfying the requirements of the regulations. These contain technical specifications and the procedure for giving notice to the local authority.

Regulation A7 sets out the way in which regulations apply to alterations and extensions and Regulation A10 covers the requirement for a person 'intending to execute work to give notice and deposit plans'.

In general, any person who proposes to put up any building – even a garden shed – or make any structural alteration or extension to a building or make any major change in its use, is required by law to notify the local authority and provide it with detailed specifications, including the planned method of construction and the materials which are to be used.

If you are thinking of a single-storey extension with a floor area of not more than 30 square metres, which is to be used exclusively as a playroom or summerhouse, and it is to be sited away from the house, then it is only necessary to notify the council offices of what you are planning without having to go to the expense of preparing detailed drawings.

If you don't want to buy a copy of the Building Regulations, costing £1 (with amendments extra) and it is not exactly bedside reading, your local council offices and the public library will let you see a copy. As I say, it is a complex document which is better understood by a professional, but it's there if you want to delve into it. Your architect or builder will point out any modifications that may have to be made to your original ideas and the local authority Building Inspector will usually be happy to give advice on the main issues.

Extending sideways

You may consider building over an existing garage; the first thing that has to be established is whether it can take the added weight. It is unlikely that it will, for the foundations will have been planned to take a single-storey construction. Here again, the professional is the only one who will be able to tell, although it may mean digging down close to the foundations to check. Before you go to these lengths, do have a word with your local Building Inspector. The obliging ones will pop round and give advice, although they're not required to! He might be able to check on records if the house was built after 1945.

The renewing of foundations and building of new walls is an expensive business but there may be no alternative. Where an extension is only needed to provide a bedroom or extra bathroom and a double-storey structure would block access to the rear of the house, the answer could be to support the rooms on reinforced brick piers. This would leave the space underneath the extension as a useful car port. For the purposes of satisfying the planning authority you would need to take into account the fact that the cubic area of the empty space beneath the room or rooms would be counted in when totting up the total area of the proposed extension.

Using the loft

Where an extension involves structural changes – and this applies particularly in the roof – then there are a number of points worth checking before you call in the experts.

Types of roofs vary between 'ideal' and 'almost impossible' and if you fall into the latter category there is obviously little point in thinking any further about converting the loft into an extra room for a bedroom or playroom.

If you walk into the garden and look up at the back or front of the house or bungalow, you'll see pretty clearly from the following list of diminishing possibilities whether or not a loft conversion is a serious contender.

Bungalow Usually perfect because of the immense space in the loft area. It will often take a suite of rooms. The bungalow featured later in this book is a perfect example of how successful this sort of operation can be, for the extension virtually doubled the size of the building.

Terraced (or town) house with gabled roof. This is fairly easy because the necessary height from first-floor ceiling to the underside of the roof is there. The ceiling of a habitable room must be at least 2·3m (7ft 6in) above the finished floor level.

Semi-detached house with a single hip roof. This does not provide too many headaches and a dormer window will give the extra overall height that is required.

Detached house with a double hip roof. That is the sort where the roof rises to two points and a ridge runs between one apex and the other. This type can be difficult and if you have not got the minimum 2·45 (8ft) from the attic joists to each apex (reduced to 2·3m (7ft 6in) with a new floor as I mentioned above) then it is advisable to forget the loft.

Detached house with a pyramid roof. As its name suggests, this is the roof that comes to a peak and unless there is more than 3–3·5m (11–12ft) from attic joists to the highest part of the roof then there is little hope of a successful and economic conversion. New joists will have to be put in to support the load of a new floor and the cold-water tank and expansion tank of a central heating system may well have to be moved to create sufficient space for the loft room.

A bungalow is usually ideal for conversion because of the immense amount of wasted loft space. It can often take a suite of rooms.

Town house with gabled roof usually has the necessary height for an attic room.

Semi-detached house with single hip roof is fairly simple to convert by adding an extra bedroom or playroom.

Detached house with a double hip roof. This can present big problems in achieving the regulation ceiling height for an attic room.

Detached house with pyramid roof. Achieving the required height may again present a problem.

The roofs of some period houses were constructed as a well: the opportunity of 'infilling' and creating a room is there but it will be an expensive job.

Planning permission

The ideas you have may need to be adapted to come into line with the requirements of the Town and Country Planning Act 1971. This states that any development of land requires planning permission and it goes on to define this as 'the carrying out of building, engineering, mining or other operations in, on, over or under land, or the making of any material change in the use of any buildings or other land'.

You can safely assume that, from a pre-planning stage, any major work will need planning permission, but that a comparatively minor extension which does not alter the external appearance of a building will not. Indeed, the 1971 Act points out that certain jobs such as work which affects only the existing interior of a building or does not 'materially affect' the façade is automatically permissible.

'Permitted development' includes the enlargement of a house by not more than 1750 cubic feet or one-tenth, whichever is the greater, and subject to a maximum of 4000 cubic feet. You must not exceed the height of the existing house nor must the extension project beyond the front – the building line.

You can work out the cubic capacity of your home by measuring the sides externally, multiplying the length by the breadth and then multiplying that figure by the height of the house. This is fine so long as the house is conveniently square or rectangular with a flat roof, but if it has all sorts of irregular corners and a variable pitched roof, then it gets more difficult; this is probably the point at which you call in an expert.

The question of external appearance is a vital one. Local authorities are strict about the finished extension blending in, not only with the existing property, but also the surrounding houses. There are always exceptions to the rule. In the vicinity of my own house in Surrey, a house has been made to blend in with an extension (rather than the other way round) by using a slate-grey cladding that makes it right out of keeping with neighbouring houses which, although varying in precise appearance, are uniform in their brick finishes. The point, therefore, is to be positive in your approach to the local authority and reasonably firm if you feel strongly enough about the proposed look of an extension.

The type and colour of bricks too can affect the ultimate success of a home extension both in terms of appearance and timing.

The house on the right of the picture has been finished in a style which is certainly not in keeping with the surrounding properties but expresses the individuality of the architect who designed it. The existing house was rendered to blend in with the extension.

Approval of your plans by the local authority may depend on blending in with matching bricks and if the house is old, then it's as well to inquire as to how and where you might get second-hand ones, perhaps from an old property being demolished somewhere in the area.

Where an addition to an existing building is to be, say, a self-contained flat for elderly parents, a preliminary check should be made with the local planning office, because they could look upon this as a change of use, from a single to a two-family dwelling, which immediately increases the density in the locality.

They could, therefore, insist that any extension should not be entirely self-contained; this could be complied with by providing a door through from the present house or building the extension without its own kitchen.

As a general rule, planning is concerned not with a householder's rights or entitlements, but what is for the public good. For instance, you are not actually entitled to a view, but your neighbour may object if an extension comes so close to his property that it cuts out the daylight or has side windows that look into his house. Each case is judged on its merits by the local authority.

Another checkpoint in your pre-planning is that if you want the extension to be clad with timber boards then it will need to be at least 6m (20ft) away from the boundary to conform with fire regulations. You'll be restricted too in the number and size of windows on the boundary wall. You cannot have a full-size window or ordinary door in that wall if the building comes within 1m (3ft) of the boundary.

Fire regulations govern the construction and siting of an extension to such a degree that the whole question of proximity to a boundary cannot be overemphasized. In lay terms, fire resistance is the length of time a wall will resist fire, and for an outside wall this must be not less than half an hour if it is sited less than 1m (3ft) from the boundary.

The regulations, like the Building Regulations generally, are not the simplest works to comprehend. While your local builder may have a general appreciation of their requirements, it is the architect or surveyor who will be more intimately versed in their finer points.

If you are living in a listed house of special historic or architectural interest, authorization will be required before you even start sketching out any ideas on paper. This official approval has to be sought through your local authority.

Conserving heat is all important, especially when it comes to the loft which is 'open' to the weather on all sides. There are various ways you can insulate not only keeping the heat where it belongs – inside – but keeping noise out. In our drawing the roof has been lined and then tiled; the walls covered with insulating board, or alternatively timber lined; the dormer windows double glazed and the floor made up of insulating boards. The point to remember about insulating an attic room is that heat rising up through the new floor from the house below can be put to good use, so it's the roof that calls for extra special attention.

Insulation

Conservation of energy is today a critical factor and rightly so because so much heat can go to waste without adequate insulation. While an existing property can be expensive to insulate, particularly for double glazing and cavity filling, an extension is comparatively cheap. Building Regulations determine a certain minimum, but you can go a long way to improving on this.

The roof, especially where it is flat, will benefit from additional insulation and noise-absorbing material. Several owners of the flat-roofed loft conversions that I've visited in recent years have complained of being kept awake at night by rain pattering incessantly on the roof or wind whistling and whining round the room. You can't do much about wind or rain but you can certainly help to deaden the noise!

Points like these are often the ones that get overlooked, even by the 'experts', and you have a decided advantage if you build them in, as it were, to the original specification.

Access

Consider, too, the question of access from the existing building. If the new room is to be used by the children for entertaining and as a playroom, you undoubtedly will not want them tramping through the living room every fifteen minutes.

Indeed, access to the double-storey extension may be easier. In my own house, the bathroom and a separate lavatory are directly above the integral garage. When I extend sideways, access to the upstairs room or rooms will be through the present lavatory which is off the landing, so the conversion will be very simple.

There are rules to observe, too, governing access. A lavatory must not open directly into a habitable room or kitchen. The exception is a bedroom or dressing room, provided it can also be reached from another way, say a landing, or there is another lavatory in the house which has access direct to a non-habitable area like a hallway or utility room.

A downstairs bathroom will have to go without a lavatory if you can only reach it through a habitable room or kitchen. The way to overcome this potential difficulty is to build a lobby between the two rooms; it may be necessary to ventilate this in-between area. Do note, incidentally, that every new habitable room has to be ventilated. Two pages of the Building Regulations are devoted to the subject. Windows must not be too close to chimneys either. If the distance is less than 2·3m (7ft 6in), then the top of the chimney must be 1m (3ft) above the top of the window.

How to get to an attic room is another important question. If it is be a workroom or study, a loft ladder may be adequate. For a habitable room, the means of access must be permanent. Safety, with ease of access and exit, are a must and no firm should construct a staircase with less than 20cm (8in) of 'going' – that is, the depth of tread you put your feet on. No minimum width of staircase is laid down, but anything less than 76cm (2ft 6in) would be rather cramped. Where the staircase goes depends on the structure of your home; it may either rise from a large landing or it may be possible to have it going up from an existing bedroom. In any case it will need a minimum headroom of 2m (6ft 6in).

Heating

Heating a new room can be done by extending an existing central system, provided, of course, that the boiler has sufficient spare capacity – and more often than not they have, certainly for one room. In the case where a radiator system is up to its limit, then individual heaters will be the answer. The simplest will be electric, although they'll probably be the most expensive to run. A gas appliance will need to be ventilated either by providing a balanced flue – that's the sort that does not require a chimney but breathes in and out through a simple wall-mounted grating – or be of sufficiently low capacity to run safely dependent only on the permanent ventilation of a large room.

The plan

When you first start thinking about a home extension it is helpful to put your thoughts on paper. Your sketch should be drawn to scale, say 1 in. to 3 ft (2·5cm to 1m), on graph paper which is large enough for you to indicate boundaries and buildings close by.

These first ideas can then be amended as you get answers to some of the points made above. You are then in a position to offer a well-thought-out plan which will not only make it abundantly clear to the builder that you mean business but will undoubtedly save a good deal of time and frustration in seeing the job through to its successful completion. First you formulate ideas. Then sketch them on paper. The next thing is to decide on what, if any, independent professional advice to obtain and at what point in the proceedings.

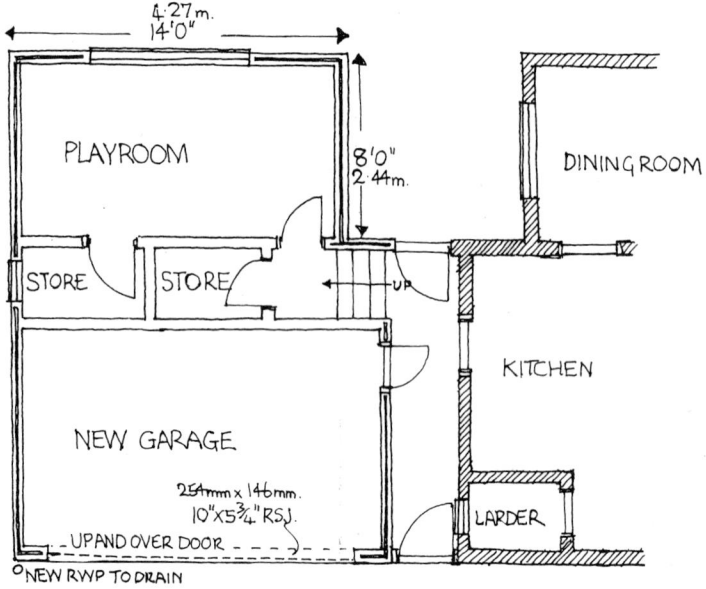

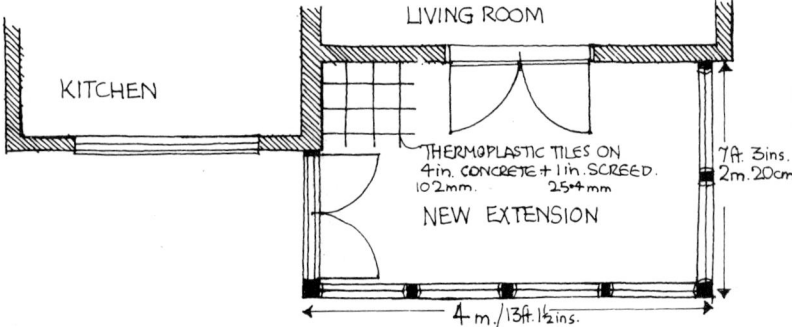

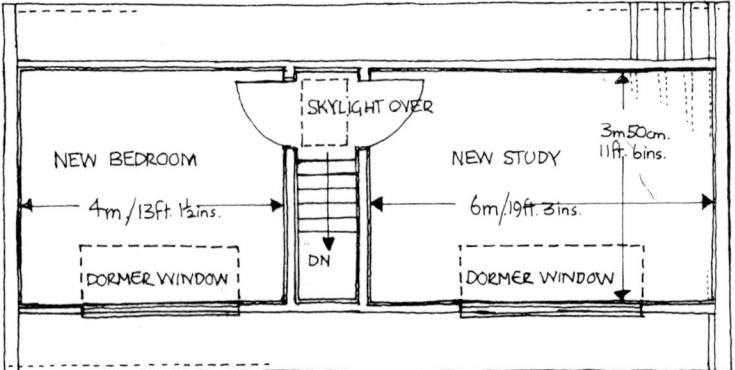

You can appoint an architect or surveyor to carry out some or all of the following tasks, although the specialist extensions firm will do most of them for you. These are the milestones in planning a home extension.

Builders and specialist firm submit estimates of cost.

Produce drawings and specification.

Get the necessary local authority approval.

Prepare a contract.

Obtain quotations from three firms.

Formally appoint a builder and complete and sign a contract.

Check on construction: notify local building inspector of stages (where a major two-storey extension is planned).

Check invoice against original estimate.

Pay final account (less 5 per cent of contract sum).

Check on possible faults.

Settle the final bill.

Some companies will ask for stage payments, in other words, financing the job for them. If there is no way of avoiding this – and it seems to be an increasing practice – then it is obviously vital to make a thorough check at each completed stage before the builder goes on to the next. This is where professional help is so essential.

Opposite are three plans for extending a home. The top one shows a custom-design addition of a playroom, new garage and useful store areas. The next plan shows how a system-built extension has been neatly fitted on at the rear of an existing living room.
The final plan is for a loft conversion, which has provided two small rooms each of which has its own dormer window. A skylight has been put over the staircase to provide extra light.

5. Accessories

Having considered what kind of extension you need to plan for, and who is going to guide you professionally there is the important consideration of what I will describe as 'the essential extras'. Rather like a motor manufacturer who leaves number plates as extras, so the home extension has a much wider variety of items that should feature on your total planning.

Such things as the type of staircase you are going to select for an attic conversion; whether it should be spiral, wrought-iron or wooden; what kind of window, dormer, pivot, single or double glazed, plain or decorative glass; what percentage of the overall budget should be spent on insulation and whether, for a ground-floor room with three outside walls, it is a worthwhile investment to have the cavity walls filled with a plastic foam to help retain heat in the house. That's just a start. Each point deserves some time to be spent on considering its relative importance. For instance, double-glazed windows are cheaper to install in a new room and while double glazing doesn't cut fuel bills substantially (which it is some-times claimed it does) it undoubtedly increases comfort and cuts down draughts; so the additional cost over single glazed windows is a reasonable 'extra'.

Stairways

Access to and from a loft room is a vital consideration. Building Regulations govern the type of staircase to be used: if it's to be a habitable room, say a bedroom, then there must be a permanent staircase. The problem here is sometimes how best to fit it in and from where it should rise. This may be from an existing landing or

Above: A spiral staircase can be expensive, but is a wonderful space-saver.
Above right: The 'disappearing' stairway that tucks away when a loft room is not in use.
Right: Conventional staircase, with quarter landing, that can rise from either an existing landing or upstairs bedroom.

UPPER FLOOR LEVEL

1·52m.
5ft. MINIMUM
AT RIGHT ANGLES

1·98 m.
6ft. 6ins. MINIMUM
VERTICALLY

LOWER FLOOR LEVEL

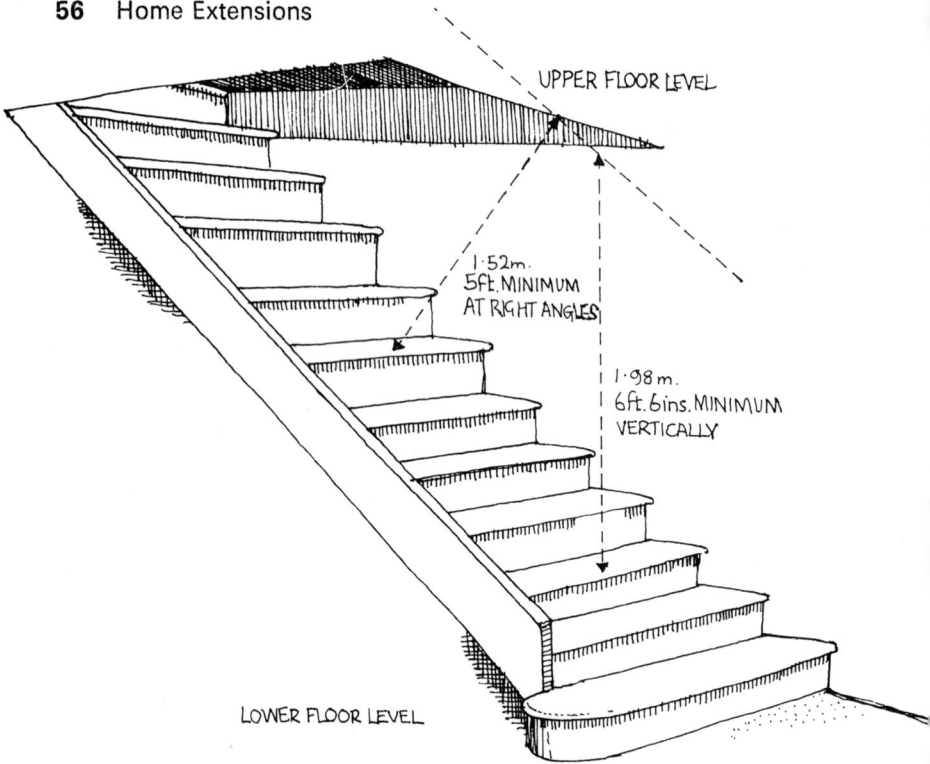

Minimum heights and distances from step to ceiling govern the
construction of stairs that lead to attic rooms. Building Regulations
stipulate the safety levels.

from an upstairs bedroom. A straight flight is the least expensive but
the necessary length more often than not rules it out. The alternative
is the same type of staircase with the addition of a quarter landing.
A further variation is two quarter landings which has the effect of
turning the stairs through 180 degrees; that way, it will fit into a
fairly narrow space.

There's yet another way and that is to have tapered steps which,
though they are a method of saving space, cost considerably more
than the more conventional stairway. A wooden or wrought-iron
spiral staircase is a real luxury but of course a perfect answer if you
are really tight on space.

Clever use of tapered steps allows the wooden staircase to a loft room to
take up the minimum amount of space.

The staircase illustrated runs up from an existing landing to a roof room and is keyed to the walls for support with the tapered steps enabling the whole stairway to turn back on itself by 180 degrees. The open design also allows for plenty of light to come through from a large, double pane casement beside the stairs which rise from the hall and a new window cut out just below the roof in the same exterior wall.

Understandably there must be a safe escape route in case of fire and the regulations require the internal stairways and associated hallways and landings in houses of three storeys (which may well be the result of a room at the top conversion) to be separated from the rest of the house by a structure with a fire resistance of half an hour.

If the room is simply to be a workroom, study or playroom, then a loft ladder is acceptable. The best type is probably that which is retractable and is made in aluminium. You may have to enlarge the loft opening to achieve a more convenient access and it's a reasonably easy job to cut away part of the joists. Again your builder is the best person to advise on this point in the initial planning stage.

Handrails Any kind of stairway needs adequate guarding of course. Indeed, those omnipotent Building Regulations again demand that certain safety precautions are included. They say that stairs have got to be guarded on each side by a wall or solid screen or a balustrade or railings at a height of not less than 84cm (2ft 9in) above the level of the actual treads. There has to be a handrail, too, on at least one side from top to bottom. Landings understandably have to be guarded to a height of at least 1m (3ft).

For the temporary stairway, like a retractable ladder, it is safer to select one that includes handrails as standard; not all do by any means.

Windows and ventilation

Windows are the naturally accepted method of ventilating a room, although there are automatic ducted systems that can be used to ventilate a lavatory or store room. If you decide, say, to add on to the sitting room at the rear of the house, the new room must maintain the required level of ventilation that every habitable room must have. It doesn't matter whether it is an adjustable louvre or a pivoted frame in the ceiling of a loft conversion, the opening must equal at least one-twentieth of the total floor area.

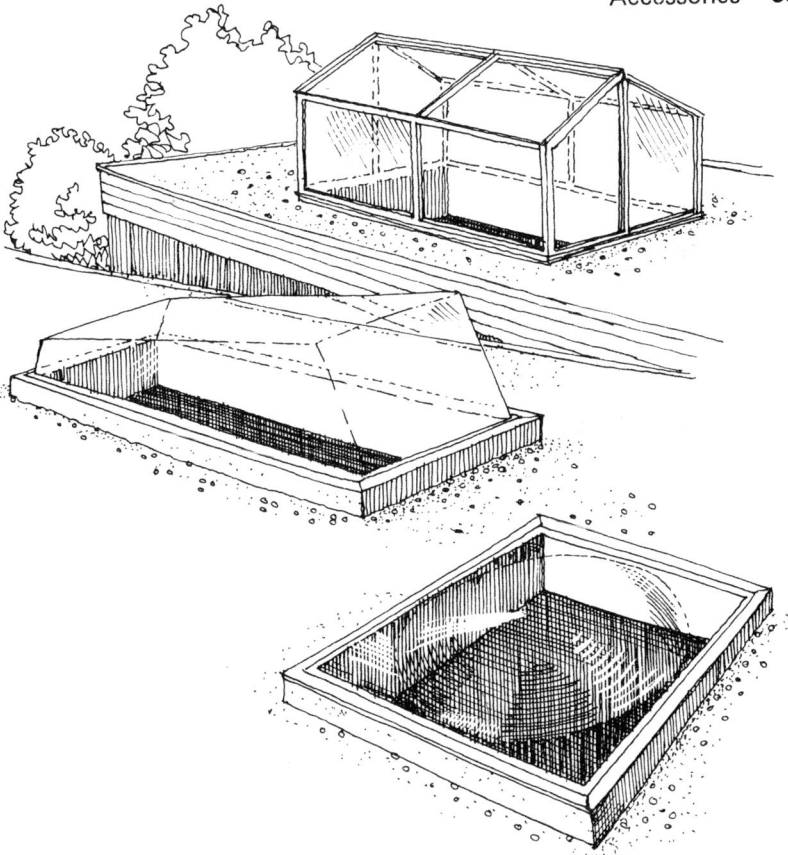

Three examples of roof windows. The top one is a greenhouse type skylight which can be installed with opening vents for ventilation. Next is a rooflight available as a non-opening light for flat roofs. At the bottom is a simple domed roof light, useful for installing in a solid-roofed ground floor extension.

French windows alone do not satisfy the regulations; there has to be an additional opening of a minimum of 100 sq cm (15 sq in) either as part of the french windows or in another window.

Space outside There are rules governing the amount of unobstructed space that must be left outside a window of a habitable room. In simplicity, this has got to be at least 3·5m (12ft). I mentioned earlier (planning check list) the point about chimneys or flues from central heating appliances being a minimum of 3 feet above a window. This applies if the top of the window is 7 feet 6 inches or less from the

MUST BE AT LEAST 3ft/ ·9lm. IF TOP OF
ADJACENT BUILDING IS LESS THAN
7ft.6ins./2·29m.OR LESS FROM CHIMNEY

MUST BE AT LEAST 3ft/ ·9lm.
WHERE CHIMNEY IS MORE
THAN 2ft./·61m. FROM RIDGE

MUST BE AT LEAST 2ft./·6lm WHERE
CHIMNEY IS 2ft/·6lm.OR LESS
FROM RIDGE

MUST BE AT LEAST 3ft./ ·9lm. IF
TOP OF A WINDOW IS 7ft.6ins./
2·29m. OR LESS FROM CHIMNEY

Strict rules apply where the conversion of an attic means that windows will come close to neighbouring chimney stacks or flue outlets from heating appliances. These particular regulations are aimed at preventing poisonous gases reaching habitable rooms or even seldom used areas such as a loft simply converted for use as a play room.

flue outlet. It's worth checking, therefore, when thinking of a room in the attic that you are not likely to have to ask your neighbour to extend his brick chimney stack a foot or two! The usual height of a stack, plus the distance between houses combined with the pitch of the roof should make this a million to one chance, but err on the side of caution.

Windows up   A dormer window or windows in an attic room will often provide that necessary extra height to meet Building Regulation needs. It can be an attractive feature of the house, too, especially when it's in the older style. The sides can be tile-hung to match in

Opposite top: Standard dormer window giving extra height in a loft room as well as the necessary amount of ventilation.
Middle: Roof window with pivoting sash gives better light than a dormer window, but can only be used where there is the regulation height inside the room.
Bottom: Roof light for flat or pitched roofs.

with the roof. A dormer window on the front of the house will need Town and Country Planning approval.

Where the required height is already there, then a pivot window, which can give more light, is a successful alternative. They are opened by means of a full width handle rail at the top of the sash and have a ventilation flap that can be used while the window itself stays closed.

Safety, particularly where the loft room is to be used by younger children, is catered for in some pivot windows with barrel bolts that secure the window at two points in a partly open position.

For the dormer window, metal bars across the width of the opening sash are a necessity.

Windows down   Plain, opaque and decorative glass are the three main choices for the windows of ground-floor extensions, and the choice is bound to be dictated by the siting of the new room. If there's any chance of being overlooked – for instance through the end or side windows of a garden room tacked on to the back of the house – then opaque or decorative panes in these windows and plain in the ones facing down the garden could well be the right mix.

Somerset Maugham once said that the Englishman's idea of luxury was to eat a cold lamb chop in a howling gale! Central heating has changed all that, even if it were true. But the new accent on energy conservation, has brought in its wake a demand for keeping the heat where it belongs, inside. Double glazing of the windows of a home extension, which will often have two or three outside walls anyway, can be a useful, although not cheap, way of keeping the room warmer in winter. (Remember that as it helps keep the cold out so it will do the same with the sun!)

The first and best choice for the windows of a new room are those which are known in the trade as 'factory sealed units'. As their name implies, they are ready-made windows with the two panes of glass sealed together with metal, alloy or plastic edging. They are thus simple to install instead of a single-glazed frame.

If you are having a brick-built room on the back of the house, double-glazed sliding patio doors are worth looking at. I've already mentioned my reservations about the claims that double glazing saves fuel costs. In addition to its value in increasing comfort, the secondary sash type has a security advantage for it can act as a deterrent to the would-be burglar, especially the 'quick in and out' sort.

Insulation

This leads rather naturally to insulation. It is reckoned that some 75 per cent of the heat generated in an uninsulated house disappears through the roof, walls, windows and cracks in the structure. Although the worst offenders are windows, the actual space they occupy puts them down in the list of priorities. It's roofs and walls that come top.

The benefits of investing in this kind of 'built-in' structural advantage are manifold. It keeps the house warmer and it helps to preserve the good decorative order of the house. Cavity walls of a brick extension can be filled with a plastic foam that is pumped in under pressure and sets to form a barrier against the cold, without preventing the cavity doing its work of stopping damp penetrating to the inner walls, but do insist on a guarantee to cover this point.

The home extension described on page 86 was only permitted by the local council after a good deal of detailed argument and assurances about insulation and damp-proofing.

Insulation of the flat roof of an extension will be built in by the contractor, working to the architect's drawings, or, in the case of the ready-to-assemble rooms, it will be a standard feature included by the manufacturer. Improvements, such as a board and felt roof, can be had as an optional extra, but you may prefer to call in a specialist roofing contractor to complete the final stages. You can always add to it though by asking for an internal lining of insulation board. Where it's a pitched roof, then glass fibre matting or pellets could be laid between the ceiling joists.

If the extra room is to be used for living accommodation then a translucent vinyl roof – popular along the many types of sunrooms – cannot be used for it will not meet the requirements for thermal insulation. It is likely to create condensation, too, in winter.

A solid flat roof of asbestos-cement sheeting can satisfy the regulations having polystyrene bonded to it to give the right level of thermal insulation.

The finishes that you can choose from for walls and ceilings depend, initially, on just what type of extension you have plumped for.

The wall linings of a ready-made room for use as a leisure room, for example, can be of 5cm (2in) fibreglass and foil-backed boards fixed to the rood or concrete exterior walls of the extension. Provision in some models is made for the fitting of wall cupboards.

The internal wall surfaces of a brick-built extension are usually

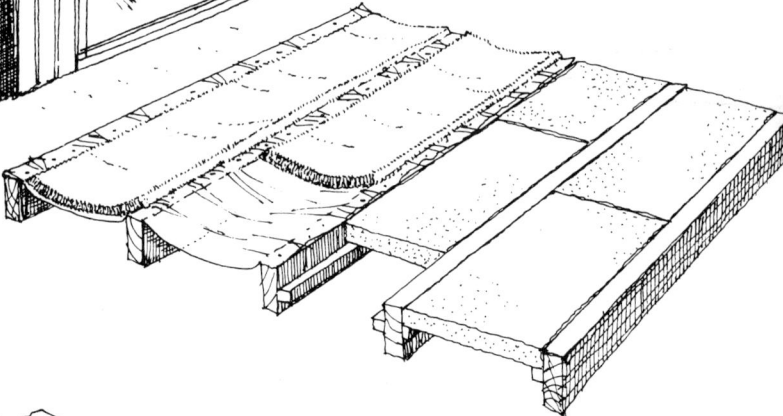

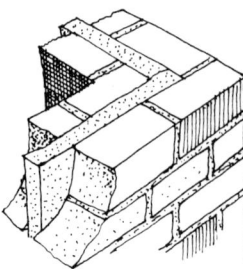
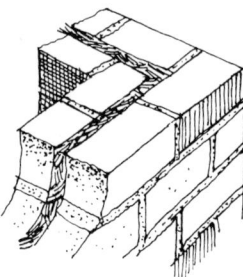
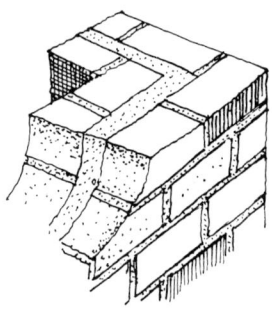

plastered; it's the simplest, certainly, and allows for easy wall-papering. The alternative to plaster is to fix timber boarding direct on to the brickwork with the use of battens across the walls. The wood must be treated with a flame-retardant liquid, unless you are restricting actual wood finish to less than half the total floor area of the room.

One of the options open to you if you should want to trim the budget to the real minimum is to elect to carry out the interior finishes yourself. For instance, with a timber frame wall, plaster-board is often fixed to the vertical and horizontal timbers to give a solid interior finish. Although this can be wallpapered without needing any further treatment, if you decide to paint it, then it is advisable first to put a coating of plaster over the boarding.

Electric power and lighting points also constitute a finishing 'extra' but unless you are a dab – and experienced – hand at fixing new points it's as well to leave it to the professionals. The provision of a suspended ceiling with inset spot lamps could be the sort of additional feature that you may feel you want to do yourself.

Whatever you finally decide to do, it is better to have too many points than too few and if the house is old and needs rewiring this could be the right time to have that job done too.

The conservation of heat is a vital consideration when planning an extension. The sketch at the top illustrates how various forms of insulation can be used to improve the ground floor extension with the use of glass fibre, expanded polystyrene and woodwool on the roof; fibre glass as an infill between the outer and inner walls and double-glazing of the windows.

The centre drawing has a choice of floor coverings including chipboard, tongued and grooved wood, cork tiles and carpet.

At the bottom are three cut-away sketches of cavity insulation. The first uses polystyrene panels, the second woodwool and the third a special foam fill, which is pumped into the cavity under pressure.

Information on all these forms of insulating materials can be obtained from Home Improvement Centres or the Building Centres in London and Glasgow.

6. Raising the Money

Can you afford it? Is it going to give you a good return on your investment of hard-won capital? These are the two most critical questions with which you will be faced having looked at the various ways for extending your home and obtained estimates from two or three builders.

It is wise, anyway, to allow for a contingency sum to cover 'unexpected extras'. Buildings costs rise, as do the prices of fixtures and fittings and a world shortage of plaster – not that that's forecast – could add a shocking amount to the final bill.

You may be drawing on capital or borrowing or perhaps combining the two: part capital in the form of a 'down payment' and part loan.

Monthly repayments on loans are one of the different forms of interest that are eligible for tax relief. There are various forms of short-term credit plans which allow for clearance of the loan within nine months, and it's as well carefully to check which scheme is going to suit your bank balance best. It might pay you to spread the loan over as long a time as possible, so that you get maximum possible tax relief, but this will, of course, mean a greater amount of interest being paid compared with a shorter repayment period. Most of the ready-made home extension manufacturers offer their own or associated credit facilities. One major company has its own finance company within the group, formed with the specific intention of 'providing an efficient and confidential service to customers'.

Part of their loan conditions include the apparently altruistic provision that they have no right of repossession, once you have signed the agreement and the extension has been delivered. Not that anyone who has gone to so much time and trouble is likely to default; nevertheless it's a comforting vote of confidence in that particular company's customers.

Typical of the contractual qualifications are that you should be over 18, a houseowner, and in full-time employment. One contract I saw recently would cover a loan to a housewife, but she would need a guarantor.

Always go through them with a fine tooth-comb and never be lulled into a state of security by a totally plausible salesman. He may well be right – but not for you!

Building societies are a natural source of credit and if you already have a mortgage, it may be possible to increase this to cover the cost of a home extension. The deeds of the property are held as security and the period of time for repayment will extend over the same period as your existing loan. A second mortgage will enable you to repay the money over a shorter period. The society will want to see plans of the extension and a copy of the relevant planning permission. Their staff valuer will also need to inspect the building at various stages of construction. Your ability to meet the increased monthly repayments to the society will also have to be proved.

While the latter half of 1975 saw a generous amount of cash flowing into building societies, priority on home loans was still being given to first-time buyers. The government have also promulgated a scheme, via the local authorities, for helping young people, so you may not, as a consequence, be quite at the top of the list for consideration. It will largely depend on the amount you wish to borrow. You could stand a better chance of borrowing money if you already save with a building society. They have a great respect for investors.

Banks are a further source of money. Barclays have a personal loan scheme and a non-secured scheme called Barclayloan which has a ceiling of £1500. The maximum repayment period is three years, but this can sometimes be extended to five years in special circumstances. The scheme is available to both customers and non-customers. Interest is 18·4 per cent and averages between 3 and 5 per cent above basic rate.

National Westminster have a special Home Improvements Loan Scheme which is designed to provide medium-term finance to owner-occupiers. In addition to the cost of structural work and improvements, the scheme can cover the cost of furnishing an extension, including fitted carpets.

Loans are granted in the range from £500 to £5000 usually over a period of one to five years at $7\frac{1}{2}$ per cent secured, or 9 per cent unsecured. Repayment is made by equal monthly instalments covering the principal sum and interest and the emphasis is usually on providing an insurance policy as security, giving life cover equal to the amount of the loan.

As to creditworthiness, the bank in general considers that a borrower's total commitment on any mortgage or home improvement loan should not exceed a quarter of their annual income.

The Midland have two basic schemes for personal loans. One has an upper limit of £1000 with repayment over two years; the second has a ceiling of £5000 with the money being paid back over five years. Interest rates vary between $8\frac{1}{2}$ per cent and 9 per cent.

Lloyds Bank offer personal loans generally repayable within five years, with interest 'negotiable' but usually up to 5 per cent over base rate.

It is most important to remember that when banks, or any company for that matter, lend money on a *flat* rate of interest, the true rate that you actually pay over a number of years is considerably higher. For example, 9 per cent flat is equivalent to 17·2 per cent over a two-year period.

Insurance companies can sometimes provide a cheaper method of finance, especially if you have had a life insurance policy out for some years and a good deal of premium income has accrued.

The loan will be worked out on the surrender value of an existing policy. They may also grant a loan on the security of property, but only if this is not held by a building society on the strength of a mortgage.

Any organization willing to lend you money will make the loan dependent on a surveyor's valuation of the property. Therefore, raising money through a building society by extending a current mortgage will be successful if the existing house, which is to be the security, has increased in value sufficiently to make the loan a good investment from the society's point of view. You will have to pay that surveyor's fee, even if his report is unfavourable and your application is turned down. The way to protect against this is to ask a local estate agent to give you an approximate valuation of the house which he will see as potential new business. You're not being dishonest: it is just a pretty foolproof method of avoiding disappointment and wasting money.

Finally, try the treasurer's department of your local council. Local authorities make loans under normal circumstances, but cuts in their levels of spending may well make an application to them for a loan a chancy business at the present time.

Improving a house by adding a bathroom as part of a single or two-storey extension at the rear can be partly financed by a House Renovation Grant. These apply to all types of homes built before October 1961, and there are various categories of grant. They are aimed at assisting owners to bring their properties to a higher standard and will not cover the cost of adding an extra bedroom.

Grants for improvement are restricted to dwellings with a rateable value of up to £300 in the Greater London area and £175 elsewhere. If you should qualify under these limits, then do note that the ceiling on costs which may be grant-aided is usually £3200 per dwelling. Where a building of three or more storeys is converted to provide flats, then the limit rises to £3700 for each flat.

The local council has the final word in the granting of aid for improvement or repairs, although grants for providing any basic amenities like a bathroom or piped water supply are a direct entitlement, so long as certain conditions are fulfilled.

The maximum amount for which you are eligible, bearing in mind that a grant will normally be 50 per cent of the total cost, is £100 for the installation of a bath or shower, £50 for a wash basin, £100 for a sink and £150 for a w.c. The eligible expense for a hot and cold water supply for a bath or shower is £140, with £70 for a basin and £90 for a sink. Grants are also available for repair and replacement up to an additional eligible expense limit of £800.

No grant is paid until the work is finished to the satisfaction of the local authority. Registered disabled people are looked after in the grants scheme with money being available for improvements in their accommodation where it is unsuitable or inadequate. You can get specific details from your local council offices.

The table below, prepared by one major home extensions company, gives an idea of just how interest rates on credit plans are worked out.

CREDIT PLAN
Deposit required 33⅓% of Cash Price

£	12 months		18 months		24 months	
Amount of loan	Rep.	Int.	Rep.	Int.	Rep.	Int.
1	0·10	0·20	0·07	0·26	0·06	0·44
2	0·20	0·40	0·14	0·52	0·11	0·64
3	0·29	0·48	0·21	0·78	0·17	1·08
4	0·39	0·68	0·28	1·04	0·22	1·28
5	0·48	0·76	0·35	1·30	0·28	1·72
10	0·96	1·52	0·69	2·42	0·55	3·20
20	1·92	3·04	1·37	4·66	1·09	6·16
30	2·88	4·56	2·05	6·90	1·63	9·12
40	3·84	6·08	2·73	9·14	2·17	12·08
50	4·80	7·60	3·41	11·38	2·71	15·04
100	9·59	15·08	6·81	22·58	5·42	30·08
200	19·17	30·04	13·62	45·16	10·84	60·16

SHORT-TERM CREDIT PLAN
No Deposit required

Cash Price £	9 monthly payments of £	Total Cost £
1	0·13	1·17
2	0·25	2·25
3	0·38	3·42
4	0·50	4·50
5	0·63	5·67
10	1·25	11·25
20	2·50	22·50
30	3·75	33·75
40	5·00	45·00
50	6·25	56·25
100	12·50	112·50
200	25·00	225·00

The above examples are only a guide.
By courtesy of Banbury Houses.

7. Advice

Consumerism, the fashionable word meaning 'the customer is still entitled to be right even if he's bemused, befuddled and bedevilled on many occasions', is meant to help us all in our search for customer satisfaction. The Consumers' Association is the most well-known and effective guardian of consumer interest. Through its publication *Which?* the Association regularly aims to advise on many areas concerning the home; it has also published useful guides on a consumer's rights in law. If you do have problems then go to one of the new Consumer Advice Centres.

Architects

The Royal Institute of British Architects at 66 Portland Place, London, W1N 4AD has a client's advisory service which will help you to find an architect. The service selects a number of architects in your area and then works out a short list for your consideration based on your initial instructions. This is a free service.

Surveyors

If you decide instead to employ your own specialists, then you can get advice on the selection of the right surveyor for the particular job from the Royal Institution of Chartered Surveyors at 12 Great George Street, London, SW1P 3AD, or in Scotland from the RICS at 7 Manor Place, Edinburgh EH3 7DN.

What do a chartered surveyor's services cost? Not all professional services can be costed in advance – but fee scales have been devised as a guide. Most of these are *ad valorem* scales, that is to say they are based on the value of the property concerned – but some are worked out on the value of time devoted to carrying out a client's instructions. The scales of charges are set out in a small handbook obtainable from the Institution.

Home Improvement Centres

Authorized Home Improvement Centres, appointed and sponsored by the National Federation of Builders and Plumbers Merchants, offer free advice on a wide range of topics. There are 500 of them throughout the country and they are checked on regularly to ensure a certain standard of service. Locally, you will recognize them under their trading name, for example, Hall & Co. of Redhill or Sankey (Yorkshire) Ltd. of Doncaster. You can obtain a full list of centres from the offices of the Federation at High Holborn House, 52–54 High Holborn, London, WC1V 6SP (Tel. 01–242 7772).

The larger Centres stock a wide range of domestic units, appliances, fittings and materials and will help in installation and by suggesting the names of builders to whom you can write for estimates for the extension work that is to be carried out.

They will also be able to assist in the first stage of planning a house improvement or conversion if you find the prospect of any form of initial planning a daunting one. The Centres often provide the setting for appliances so that you can more readily appreciate how they might appear in your own home. Each Centre tends to specialize in one particular aspect of home improvements so that, taking the two firms instanced above, Hall & Co. of Redhill places special emphasis on bathrooms, kitchens, central heating and home and garden construction while Sankey (Yorkshire) Ltd. specializes in central heating and fireplaces.

Double glazing is also often included as an expert service and there's a good opportunity of seeing the different types available.

The advantage that such Centres as these have is in the fact that they are all in the same specialized business – and in most instances have been for many years – and their co-ordinated strength under the auspices of the Federation offers something of a guarantee of reliable service which is an increasingly valuable and all too rare commodity today. The Federation estimates that it represents 95 per cent of all builders' merchants in this country. You do have to look for the word 'Authorized' though, because in some towns Home Improvement Centres have sprung up outside the Federation's control. Do check this point when looking through the Yellow Pages, which is the other source of local addresses.

Home Extension Building Association

An organization which claims to give you 'the best of both worlds' – the personal attention of a recommended local builder plus the backing of a counselling service – is the Home Extension Building Association. Basically the Association is a window dressing for a group of 'financially independent contractors within the building trade' which simply means companies outside the big groups. It is the brainchild of a man who runs his own firm, Home Extension Consultancy, from Weybridge in Surrey and who aims to take advantage of the boom in home extensions. The Association claims to save you money by providing a total service covering surveying, planning, design, negotiations with the local authority, raising finance, advising on possible tax concessions [sic], plumbing, insulation and the final construction work. The Association anticipated a membership of 1000 by the end of March 1976, concentrated in the southeast of England. Many members – not apparently all – belong to professional organizations such as the Federation of Master Builders. Any builder who applies for membership of the HEBA has to meet certain standards in terms of quality of work and financial stability and, in the event of a complaint, the Association 'offers direct access to their arbitration facilities to provide a speedy, satisfactory settlement'.

If you want to seek this Association's advice, then write to The Home Extension Building Association at Queen's House, 28 Queen's Road, Weybridge, Surrey.

Double glazing

Inquiries about double glazing can be directed to the Double Glazing Advisory Service set up by the Insulation Glazing Association and operating from its offices at 6 Mount Row, London, W1Y 6DY. Members of the Association are companies who make or install double glazing of all forms, including sealed units and patio doors. There are more than 150 members up and down the country and the Advisory Service will give you the name and address of your nearest IGA member company.

Again, in the unhappy event of any dissatisfaction, there is recourse through the Association whose role it is to ensure that complaints are considered and corrected as swiftly as possible.

Seeing the goods

Places where you can see examples of domestic appliances and obtain general leaflets on materials for home improvements and fitting out a new home extension are the Design Centres in London's Haymarket and at 72 St Vincent Street, Glasgow C2, and Building Centres in Store Street, London, just off Tottenham Court Road, also at 6 Newton Terrace, Glasgow G3 7PF.

Most of the major home extension manufacturers have show sites displaying their ranges of sun rooms, day and leisure extensions and, helpfully, nearly all sites are open on Sundays. A visit to one of these display centres will enable you to get first-hand advice and see just what you might be buying.

There is no corporate advisory service or centre for the firms making already assembled rooms, but a glance through the advertisements in local and national papers will provide you with a selection list.

Advice and recommendation from a friend or neighbour who has recently extended his home will serve you in good stead when it comes to taking the final decision.

Case Histories

8. Playing the Waiting Game

In January 1972 planning permission was given to a young family to extend their Surrey farmhouse. By October that year the work was still uncompleted. Somewhere along the way nearly four months had been lost – time spent in waiting.

Most of the delays were because the owners ordered non-standard equipment such as a 6ft bath, standard kitchen units that had to be tailored to fit the existing kitchen and a stable-type door to be hung between there and a new playroom-cum-breakfast room.

What it illustrates very forcibly is that if, like many other people today, you are making more room in your present home rather than move, then you may well have to play the waiting game.

The farmhouse is sixty years old and built in a hard engineering brick – the sort that was used in constructing the old railway bridges. Three to four weeks was quoted for delivery of matching bricks; the actual time taken was seven to eight weeks. Then they came in metric sizes so the builder had to double the thickness of cement between courses in order to line up each course with the existing brickwork.

The next delay was because the family ordered a 6ft bath; this took twice as long to get as a 5ft 6in model. The side and end panels for the bath arrived weeks after the bath had been fitted.

The 'lead times' quoted by manufacturers – this is the estimated length of time it takes from the receipt of an order to its actual delivery – changed constantly. The reasons given ranged from 'industrial action' to 'unprecedented demand'. This caused further problems in planning and timing the various stages of work for the plumbers and electricians.

Specially made fitted cupboards for the new master bedroom in the extension were finished behind schedule after several joiners had quoted up to a three months' waiting list. Then the family had to pay extra to speed up the job.

New kitchen units came seven weeks later than promised.

There was even a delay in obtaining a towel rail – and then they did not get the one they really wanted.

Cork vinyl tiles were chosen for the playroom and the concrete screed had to be given six weeks to dry out naturally before the tiles could be laid.

Where the family did manage to save was in the cost of the extension. They employed a small local builder known personally by the owner who was certain of a top-rate job – and he was right. The end result is a great success and the extension has probably doubled its value already in the overall market price of the house.

No architect was employed. Instead, the owner briefed a draughtsman who then prepared plans for £30 and obtained the necessary planning consent from the local authority.

With three-year-old and one-year-old boys at home, the owner's wife had to be blessed with the patience of a saint, and, as she says, 'It really opened our eyes to the incredible delays that can occur in what should be a fairly straightforward building job.'

Their extension, which added a large bedroom and bathroom *en suite* on the first floor and a 23 × 13ft downstairs room, is also a good example of well-thought-out planning.

They bought the farmhouse some eight years ago and found living in the heart of Surrey not only far enough away from advancing suburbia for peace, but convenient for the husband who worked in South London. But they needed more space, not just then but for the future. The extension was therefore planned as an investment, and at an approximate £7 per square foot it had to be.

The cork floor tiles are for easy cleaning, particularly while the children are young, and make an excellent dance floor surface for when they grow up. Lighting is from overhead points, but there is also provision for wall lights, so that the room is as flexible in its use as possible.

In the owner's words the new extension has 'transformed an ordinary house into a permanent home'.

The addition of this two-storey extension was a major investment for a young family. (*A. Francis*)

9. Room with a View

Growing room was the basic reason that led a young Kent couple, who liked the three-bedroom, detached house they had bought earlier but needed growing room to share with three boisterous lads, to add on another bedroom.

It meant building over an existing driveway leading to a detached garage standing just beyond the rear of the house, so they've ended up with a second garage at each end of which are up-and-over doors. The extension has given the family a room 16ft long and 8ft 6in wide. They would have liked a second bathroom but this would have meant problems of ventilating an existing upstairs lavatory and that, combined with the total cost, decided them against it.

As it was, drains and main services had to be resited and new soakaways provided.

An architect friend designed the extension to their own very clear brief and the balcony, which gives such an attractive exterior façade to the room, was the husband's inspiration.

The house is in the London Borough of Bromley and the local authority maintained strict control as the job progressed, from the foundations for the supporting piers, upwards.

Fire regulations come very strictly into force with home extensions, particularly where they involve upstairs rooms and here the roof of the newly-created garage and floor of the extension was lined with asbestos. All external pipes were closed in with asbestos.

The gabling of the roof cost a little more but the family agree that the way the whole job now blends in with the original house was well worth the extra money.

A thoughtful and practical feature was the exclusion of guttering from the side that butts on to their neighbours' property: it means

The balcony was a bonus of this extra bedroom built above the garage.

that there is little, if any, maintenance which would have necessitated running ladders up from next door.

One word of caution. The job took nearly twice as long as had been estimated – a total of eight months – and the family feel that a time clause is a necessary part of the contract with the builder.

Their advice is also to ensure that you employ a builder with his own regular team of specialists, such as the plumber and brickie and therefore doesn't sub-contract: that, and the shortage of materials are the two main causes of delay.

The extension to their home cost £2000 in 1974 and has added considerably more than that to the re-sale value of the property.

10. Growing Up

The reasons for wanting to extend a house or bungalow are varied, but the most common is the young family that needs extra space in which to grow.

There were two reasons for the loft conversion, shown here, which was carried out by Crescourt. The first was the fact that the two children in the family, a boy and girl growing into their 'teens, had reached an age where separate bedrooms were desirable. The second requirement was for a music room in which the family could practise their instruments undisturbed.

Three rooms were therefore created out of what was fundamentally wasted space at a cost of £1400 in 1973. The whole job took fourteen days to complete, excluding the time spent on preparing plans and getting local authority approval.

The large area of 'wasted' space in the attic of this bungalow was converted into three rooms.

Interior view of one of the three bedrooms built into the roof of a conventional detached bungalow in a loft conversion.

11. A Century Not Out

Just over 100 years ago it was built as a farm-worker's house – two rooms up and two down – a quarter of a mile from Brighton seafront. Today it stands amid rows of terraced dwellings as a reminder that with a little thought, a dash of creativity and a radical yet seasoned approach to planning, you can produce a recipe for an enchantingly modernized family home – now three rooms up, plus two bathrooms, and three rooms down.

The house began its second life two years ago when it was bought by a woman writer. She sold her London home and this financed the extension that was planned to give the Brighton house a new lease of life.

The new owner's brief to her consulting engineer was for an additional living room downstairs, an extra bedroom and a modern kitchen: the existing one was little more than a glorified scullery.

Plans were drawn up for a two-storey extension at the back of the house, so that the building is now L-shaped, with the new living room at the rear opening on to a paved terrace and patio garden that fills up the rest of the plot.

Two of the major considerations in planning the job were first, it meant going up to the rear boundary with a two-storey high structure and secondly, windows had to be installed with a practical yet architecturally pleasing elevation where there had been none before. The back of the old house had been blind.

On the second problem, the consulting engineer worked over several weeks planning a whole variety of different elevations before producing a successful solution. This included double-glazed sliding doors opening out on to the terrace from the new living room.

The local authority, faced with a conglomeration of extensions that

The front of the house hides the two-storey addition at the rear.

Looking through
from the existing
living room to the
new extension.

are largely hidden behind so many houses in the same locality, realistically raised no objection to the plans for a two-storey structure.

It was the way it was to be built that really created a precedent. The idea was to use solid 8in blocks instead of bricks. These made for a quicker construction and, so the engineer claims, a stronger finish.

It took a lot of 'lobbying', provision of detailed calculations and assurances about damp-proofing and insulation to persuade the local authority to give its consent. But it did and work began in December 1974.

The walls are made up of the 8in blocks which have been rendered. On the inside is a polythene lining or vapour barrier; on top of this is a 1in layer of glass fibre and finally comes plaster-board.

The consulting engineer describes the construction as 'unorthodox', but maintains that it was ideally suitable for this extension. Was it cheaper? His simple answer was 'No', but this was compensated for by the high insulation value of the walls, more than double that of conventional brickwork, he says.

Another feature of the extension is the way it drops down two steps from the ground-floor study or old living room which is just enough to give that effect of another room without the need for half walls or screens. And this came about mainly because of the desire not to interfere with the existing slate roof of the house. So the roof for the extension had to be started at a lower level.

The house, which is all-electric, now has full heating. This comes from a warm air unit built into a central position: this one is tucked away in a cupboard flush with the wall between the old and new part of the house on the ground floor. Concealed ducts carry the heat into all main rooms. An attractive archway between the ground floor living rooms houses the ducts in that part of the house.

The extension gives a living room of over 16ft × 13ft 3in added on the old living room which is 14ft 9in long but several feet narrower.

There is a combined dining room and kitchen with a peninsular unit providing a screen between the eating and working areas. Oven, hot-plates, refrigerator and dishwasher have all been built in. The owner had the peninsular unit extended cleverly to provide a cupboard for storing the ironing board, vacuum cleaner and cleaning materials. The back door leads out to a small covered area with a garden store and meter cupboards and there is car parking space for one car.

Upstairs, the old bathroom has been re-equipped and given a toilet (the only one before was outside the kitchen) and the main bedroom, which is the upper storey of the extension, has its own bathroom.

The position of the house, sitting as it does on rising ground, means it gets plenty of sun and the owner planned for a solar heater on the roof to provide domestic hot water which at present is catered for by an immersion heater.

The extension took about six weeks more than the estimated total of twenty weeks and the total cost, including all the work of refitting the existing bathroom, replacing a lot of the old woodwork, re-plastering walls and fitting out with cupboards came to over £11 000. Brighton council gave a grant of £1000. The original house cost £9000 which gives a total investment of £19 000 for a 'three-bedroom, desirable residence on the South Coast'.

12. Sitting Pretty

Copper Hall Cottage sits snugly and discreetly close by the high
street of a bustling Surrey village – a charming reminder and local
landmark of an age when stabling for the family horses was the
natural extension to a country home.

The cottage, a scheduled property, was built around 1770 as a
hayloft and stable for the house which adjoins it, Copper Hall. It was
converted some long time back into a small, three-bedroom home.

Of its long history we know little, but in 1962 a young couple,
recently married, moved in. The lease of Copper Hall was purchased
in 1967 when the owner moved and the first stage of a major home
extension was begun in 1969. (Cost £3000 including central heating.)

The first extension, which added a downstairs room plus bedroom
and boxroom upstairs, was created without a brick being laid.

It was achieved by taking over the rooms from Copper Hall. The
newly acquired area has been converted into a bedroom with its own
bathroom, an airing cupboard and landing upstairs and a room for
central heating boiler, together with a new sitting room, downstairs.

The staircase proved to be the greatest challenge to the ingenuity
of the owners. The husband, a quantity surveyor, had a head start
though on solving how it should go up from the hall, turning left to
the new rooms acquired from next door and right to the existing
bedrooms and bathroom. The problem was how and where to create
the necessary turns and leave sufficient room to get past the bottom

Matching of the brick soldier arches over the windows of this
kitchen/breakfast room extension and the clever use of rendering make
it hard to appreciate that this is a comparatively recent addition to a very
much older house.

The hallway and stairs
that now lead through
to the addition to the
house that gives an
extra living room,
bedroom and boxroom.

of the stairs through to the new rooms downstairs which had been taken in to the cottage after knocking down a partition wall.

By turning the staircase entry sideways, instead of facing the front door as it did, it was possible to put in a quarter landing for the turn upwards to a second quarter landing from which two sets of stairs run, one left and one right.

The windows have all been standardized and brick soldier arches matched on the kitchen/breakfast room extension.

The effect is not only practical but aesthetically pleasing and in character with the rest of the cottage.

1973 saw the second stage, which cost £3000, including work on equipping the new kitchen and building a detached garage. The existing kitchen was extended to make a very efficient galley-style area by the addition of a new room on the side of the cottage which provided a large breakfast area as a bonus.

Today Copper Hall Cottage is a delightfully modernized home with four bedrooms, two bathrooms, a large lounge and dining room, with an $18 \times 15 \times 7$ft L-shaped kitchen, breakfast room and laundry room.

13. Enlarging on Good Ideas

'Two-storey extension to the side of the house' was the simple title given to the detailed drawings for a major addition to the Cheam, Surrey, home of a company director and his wife and their three young daughters.

The extension has transformed the house into an outstanding property which has now established its own commanding position on the half-acre plot, by the addition of a 30 × 30ft sun lounge as a first-stage extension in 1970 and, more recently, the building on of three extra bedrooms, a second bathroom and a games room behind the garage. The original garage was partly demolished so that it could be enlarged and incorporated into the two-storey integral extension and one of the existing bedrooms was made bigger to link up with the new addition to the house.

The house was built in 1967 as a four bedroom property with bathroom, study, living and dining rooms, kitchen and utility room. Even then it was an imposing dwelling with the particular advantage of having sufficient land around it to enable the owners to plan a much larger home for their growing family. They also enjoy entertaining and the sun/garden room which leads directly from living and dining rooms is ideal for this. It is a splendid room with a great deal of natural light which is boosted by the installation of two skylights in the ceiling close to where the room joins up with the original house. This obviates any dark, shadowy spots at this end of the room during the day. The usual thoughts of moving to find a bigger house crossed the minds of the owners, but they felt that the desirability of the plot and their social ties outweighed these.

Some very useful 'extras' have been included in the extension. Downstairs, there's a small store room between the garage and the

The sheer scope of this extension indicates the major investment that was made. Additions have been made to the right and behind the original property.

playroom; upstairs, a box room provides handy storage space between the parents' and children's bedrooms.

The finished extension has been successfully blended in with the brickwork and style of the original house with design, planning and building being carried out by a firm that prides itself, rightly so, on this special aspect of its work, Anthony Purser Associates of Richmond, Surrey.

Looking through from the new sun lounge to the living and dining rooms of the old part of the house.

14. Character Reference

Cumbrae Lodge originally stood as a small, 1930s cottage which was the lodge to a large manor house in the village of Long Ditton in Surrey. Like so many houses of its type, it lacked the comforts and convenience of modern amenities; the kitchen consisted of a tiled, lean-to arrangement at the back of the house and the bathroom was off the hall, just around the corner, in fact, from the front door.

The Lodge, although basically small, offered exciting possibilities, however, and the land around it meant that there was ample room for extension and modernization. The owners, recently married, moved into the house in 1955 and five years later carried out the first stage of a major home extension.

A new kitchen, 14ft square, was built to replace the existing one and a new bathroom and separate lavatory was put above it in a two-storey construction. The old bathroom now doubles as a laundry room.

In 1966, the second stage was begun. This called for drastic alterations and provided an extension that virtually doubled the size of the original house. A new hall was built as a link between the old house and the new addition. The front door, which had previously been on the side of the house, was turned to face forwards and to feature as a centrepiece to what was to be a double-fronted property. The extension comprised a 24 × 15ft drawing room going from front to back, with an attractive bay window added to match the other side of the house. Above this large living room were built two further bedrooms, one with a bathroom *en suite*.

A new hall and front entrance (centre) link the original cottage, right, to the extension, left.

The rear view shows more clearly where the new joins the old just to the left of the upper window let into the roof.

Cumbrae Lodge today stands as a character property with five bedrooms (with one of the smaller rooms being used as a study/dressing room by the husband), three reception rooms with the former living room serving, in the charmingly pertinent description of the owner, a publisher, as 'an adolescents' retiring room'! In addition, there is now a downstairs cloakroom, part of the original bathroom.

When the manor house was pulled down to make way for a small development of architect-designed detached houses, the owners of the Lodge took the opportunity to purchase some additional land; in fact, part of the former driveway up to the main house from the road. This then enabled them to build a double garage on to the side of the new living and bedroom extension.

The Lodge now stands in the virtual centre of a half-acre plot and because the extensions have been largely carried out to the side of the property and over new ground, the overall size of the rear garden has been retained in its original layout.

The archway is where the front door used to be. It now leads to the new sitting room which runs from front to rear of the house.

15. Adding the Family Touch

If there's one person who should know how to ensure that he gets a first-rate extension, it is the managing director of a specialist firm designing and building them. I asked Tony Longstaffe of Sunnyside Home Extensions to describe how he developed his own 'run-of-the-mill' four bedroom new house in Hertfordshire into a most attractive residence which does justice to its half-acre woodland setting – one of the main reasons for buying the house four years ago.

The Need

Larger house
Identifiable community
First-class residential area
Garden of at least quarter of an acre
Four bedrooms
Two bathrooms
Study
Spare reception room for playroom
Architectural interest

Such houses are not common and you generally find *either* nice new houses in a very high density area and very expensive, *or* older houses, which are too private, very expensive or getting run down.

Solution

A small development of new houses was announced by a speculative builder in a very attractive and exclusive area of Hertfordshire. Plots ranged from one-fifth to half an acre set in natural woodland.

Playroom for the children and a small laundry room have been created out of the single-storey extension (left).

The house selected had basic style and lent itself particularly well to extension, being in a half-acre plot and tucked back from the main grouping of houses.

The house had:

Four bedrooms
One odd-shaped bathroom
One odd-shaped shower room
Adequate kitchen
12 × 30ft living room
Enormous double garage

To the left of the dotted
line is part of the
two-storey addition
that has enlarged the
living room.

Development

Phase One A 12 × 24ft single-storey extension was built on behind the garage and joining on to the kitchen back porch. The room was divided with a wall of cupboards to provide a small laundry room and a 12 × 16ft playroom for the children well away from the main reception area of the house.

Phase Two The bathroom was transformed into a compact but more practical and attractive room by simply turning the bath through 90 degrees. The shower room was enlarged to take in an unwanted adjacent wardrobe and this made sufficient room to fit a bath; much better for visiting friends with young children.

Phase Three This phase involved enlarging and up-grading the reception area by building a double-storey side extension which enlarged the lounge to 18ft wide, and provided a most useful box-room off the master bedroom above.

Natural brick and timber were extensively introduced for the internal finishings and an ultra-modern staircase incorporated. An extra door was fitted to the top of the stairs to increase the sound-proofing.

The entire house was double glazed and 4in of glassfibre blanket laid in the roof.

Phase Four This involved enlarging the kitchen to make space for an eating area. It was achieved by taking a corner of the over-generous garage and resiting the downstairs cloakroom and w.c. there and knocking the existing w.c area into the kitchen.

Cost

House Purchase	£18 000
Alterations	£8 000
Total	£26 000

The probable cost of an equivalent bespoke new house on a half-acre individual plot in 1971 was £35 000.

16. Putting on the Style

Practically all the really successful extensions today have been built from necessity.

A superb example of this is a bungalow fifteen miles out of London in Surrey where the property has doubled in size, with all but one of the new rooms created out of an enormous roof space.

The original structure was an attractive two-bedroom bungalow with living room, dining hall, kitchen and bathroom, all on the ground floor. It was bought seven years ago by the parents of Mrs Jean Higgs who, with her husband, Colonel Denis Higgs, and their family, lived close by. When Mrs Higgs's father died, it was decided that rather than her mother moving into their house, with its upstairs bedrooms and bathroom, the Higgs should move to the bungalow.

A local architect, E. R. Fread, was commissioned to plan a major extension which did, in effect, provide two homes in one. His scheme made use of a loft area of some 1000 square feet which had the additional advantage of a high-pitched roof. Into this were set five dormer windows – one each for two bedrooms, a dressing room-cum-study, bathroom and a landing. Two of the new windows are at the front of the bungalow (more practically described as a house now) one on the side elevation and two at the rear. The two bedrooms and Colonel Higgs's dressing room-cum-study have their own built-in cupboards.

There is also another room 12 × 10ft, at one end of the 22ft long landing, which at the moment is a store room, but could quite easily be converted into a third bedroom upstairs by adding a window. Access to the rooms above is from an easy-rising staircase in the dining hall.

This magnificent bungalow was doubled in size by building a whole new suite of rooms in the loft.

The whole extension was completed with the addition of a 21ft long by 12ft wide living/games room on the ground floor, with a door leading to it from the existing living room.

There were no special problems confronting the architect. Extra roof supports had to be put in and existing ones strengthened. Timbers were added above the ground floor for the extra rooms.

Windows, tiling and brickwork were carefully selected to keep in character with the rest of the property and this has been done so well that looking at the house it was impossible for me to tell that it had not, in fact, been planned that way from the very beginning.

Central heating was installed at the same time throughout the entire house and the boiler was placed in a former coal store adjoining the kitchen.

The original bungalow cost £15 000 in 1968; £6000 was spent on extending it, and today its estimated market value is £50 000.

The stairs to the converted attic lead up from the entrance hall/dining room (right).

17. Rooms in Space

The original house probably dates from 1860–70. The first entry in the land register, dated 1898, makes reference to the Earl of Lovelace, from whom the road takes its name.

The original building seems to have consisted of two rooms on the ground floor and an attic room in the transverse gabled roof which is approached from below by means of a steep winding staircase.

The floor of the larger ground-floor room is on two levels with two steps up for the full width of the room. Projecting wall piers give the impression of a proscenium to a platform, as though it were used for small meetings, but there is no evidence of this. The staircase which is in one corner of the 'platform' has been retained, and a new window inserted in the wall to improve the daylighting in this corner.

The second room – now used as a bedroom – is full height, with the ceiling following the line of the rafters, and is lit by a window high in the end gable.

At some later date a flat-roofed extension was added at the rear, housing a bathroom, a w.c., and a small room used by the present owner as a study. A conservatory was also added across the remaining width at the back, and this was subsequently given a built-up felt and boarded flat roof and converted into a sitting room.

A projecting front porch and lean-to kitchen at the side were further additions.

The present owner decided to extend the property to provide an additional bedroom and bathroom to replace the inadequate kitchen and incorporate a garage to accommodate two cars.

The new kitchen is at the rear and to one side of the existing dining room, with access between, through a lobby from which a new

The extension to this 19th-century Coach House, has provided a double garage with new kitchen behind and rooms in the roof space.

staircase leads to the new bedroom and bathroom over. A passageway has been formed through the existing roof space to the original attic bedroom.

Externally, materials have been used to harmonize with the existing building. The road frontage has yellow stock brickwork and the roof is slated with a dormer window inserted to light the access passageway. The garden elevation is faced with weatherboarding stained a deep brown, and tile hanging. From the ridge line backwards the roof is flat, with built-up felt on boarding.

18. Share and Share Alike

Share and share alike was the neighbourly approach adopted to home extensions by two enterprising couples living at Saffron Walden in Essex.

Mr and Mrs Edward Scott and Mr and Mrs Dennis Bow living in semi-detached houses claim to have scored one for the record books; they are sharing one large Blacknell extension bought from pooled resources and erected during six evenings' work. Privacy is maintained with a party wall that divides the extension in two. The Scotts use their part as a lounge and the Bows have made theirs into a recreation room, mainly for billiards. The extension is built with a new material called Timberplast, a vinyl-covered wood.

The secret behind this resourceful partnership is that Mr Scott and Mr Bow are in business together.

The reward for neighbourliness is this large ready-made extension which has been equally divided between two families.

19. Two-storey Transformation

Billing Street, Chelsea, a short throw-in from Stamford Bridge football ground, is a double row of terraced Victorian houses built around 1838. From the front they look meekly small and it is not until you get a back view that you realize they conceal a whole variety of ways of extending a home. Ideas range from a toilet on stilts, a room some 6ft square built at first-floor level on brick pillars, to a two-storey addition that has made one house more than half as big again.

The transformation is remarkable. From a two-bedroom house with a through living room on the ground floor and kitchen, dining room and bathroom in the basement, there has grown a four-bedroom house with two bathrooms, playroom, living/dining room, kitchen and workroom. The whole extension measures $12 \times 14\frac{1}{2}$ft on two floors and out of this has come a combined living/dining room (taking up part of the former eating area) and a playroom and workroom. The original living room, running from front to back of the house, has been divided into two bedrooms.

The owner, a London barrister, lives there with his family – a wife and two boys, aged 13 and 12. His reasons for staying put, having bought the house thirteen years ago, was proximity to the centre of London and the fact that moving house was an unnecessary and expensive business. That goes for the majority of people who are today planning a home extension.

The cost of the Billing Street extension was £2500 and this only covered the constructional work of laying a foundation and building the two-storey framework. The finishing work was all done by the owner. The sad footnote to this is that with the plans for a London ring-road, the house is now next door but one to an eight lane

A garden patio gives the finishing touch to this two-storey extension of a
terraced house.

motorway! That's something that is impossible to guard against,
but it does emphasize the need for researching local planning
developments when thinking of extending your home.

20. The French Line

The 'commuter belt' usually spreads anything up to twenty miles outside a major city; in that wide area there is a concentration of property in which those people tied to a commercial centre have made their homes. Thoughts of a home life in the country are scotched by the increasing cost of travel and the growing frustration caused by jam-packed roads. These factors have contributed to the largely static position of urban property and life.

They were certainly the reasons why a member of the French Railways staff, working in London, decided to stay put and extend to give himself and his family that vital growing room.

They have four children and the small, semi-detached house they moved to from a flat fifteen years ago simply could not comfortably contain high-spirited children, the eldest son of whom was indicating, understandably, that he wanted his own 'den'.

The father had his job in London and the children were at the *Lycée-Français* in Kensington; so cost was a dominant factor. As he told me: 'Time is money, too.'

They decided that the best solution was to make use of the wasted area in the loft and to convert it into an additional bedroom for their eldest son. The result is a great success providing not only an extra bedroom but a large area which is used for study and leisure.

The staircase which approaches it is cleverly concealed with a solid door, wall-papered to match the surrounding walls, so that it really looks very much like a hideaway.

The conversion of an attic in this semi-detached house has given an eldest son his own room.

The construction of the staircase has taken a small amount of space out of the front and rear bedrooms, but not enough to reduce the actual 'working area'.

The owner's one regret is that he had hoped to have a spiral staircase which would have given even more space in the loft room, but he was advised to the contrary by the builders who constructed the room five years ago. The cost was then £900.

The house has now been further extended with the addition of a large utility room leading directly off the kitchen and behind a single garage at the side of the house. This room has its own separate shower, toilet, sink and position for the washing machine.

Wallpapered door concealing a staircase leading to a room in the loft.

Index

Compiled by Gordon Robinson

Italic figures indicate illustrations